RED EYES WHITEWALLS & BLUE SMOKE

speck press

denver

Published by: *speck press*, speckpress.com

ISBN 0-9725776-3-7, ISBN13 978-0-9725776-3-2

Individual image copyrights appear on page 143.

Book layout and design by CPG
corvuspublishinggroup.com

Printed and bound in Canada

10 9 8 7 6 5 4 3 2 1

For Kilgore and Ella, the coolest scooter kids ever.

The best and most supportive friends I have I met through scootering. So many are to thank for this book. First, my best friend—my beautiful wife, Cindy, with whom my scootering life, and my adult life, began. To my brothers, Pete and Dan, who stand beside me; my mother who showed me where to draw the line; my father who gave me my compass; my Grandma Teed who's faith in me is boundless. Special thanks to Adam Baker, my scootering sponsor, business partner, and confidante. Shouts out to all of Denver City Denver; especially Phil L. Denver, Pamiel, Cho, Coolie, the Smiths, the Mega-Cs, Biggus Davus, ACE S/C, the Flying Elephants, the BRSC, JKSC (Nabu Chapter), Secret Servix, the Scarabs, and the Pub Scouts S/C (R.I.P.). Thanks also to Dana in the SLC who is the kind of guy you ought to know; Mike Mcwilliams of Peak S/C who anchored the Colorado scene for so long; and Peter Crowl who first let me ride a Vespa. Thanks to Doyne and Darleen Bruner introducing me to the motorcycle business. I must also thank my mentor and dear friend, Phillip McCaleb, for his Genuine love and support.

Acknowledgements

Pamela Anderson, Paul Andrews, Mitch Armstrong, Adam Baker, Michelle Baldwin, Brett Barker, Fabio Balerini, Ryan Basile, Rick Blazich, Doyne Bruner, Darleen Bruner, Aimee Buchwald, Pabst Blue Ribbon Beer, Colin Catel, Scott Chain, Jon Cho, Alex Cohn, Peter Crowl, Tim Deutsch, Tom Drake, Giancarlo Fantappie, Jeremy Foote, Dustin Gabel, Randolph Garner, Nelson Guanipa, Al Guillion, Barry Gwinn, Patrick Hale, Eric Halladay, Thomas Heath, Bob Herbers, Susan Hill Newton, Chrissy Hyatt, Andrew Hyder, Chris Irvin, Mark Jurus, Jaimie Katz, Roger Kilgore, Carol Kilgore, Judy Kilgore, Nick Kofski, Pavlos M. Kolovos, Al Kolvites, Missy Kroge, Jon Kroge, Chelsea Lahmers, Peter Laitmon, Moped Larry, Derek Lawrence, Phil Lombardo, Matt Lombardo, Mike Loop, Joel Martin, Chris Martinez, Phil McCaleb, Margaret McCullough, Alex McKenzie, Jeremy McMinn, Mike Mcwilliams, Matt Megyesi, Michael Menaker, Vince Mross, Wade Parker, Alex Pelzel, Jack Penley, J. B. Penner, Simon Penner, Chris Peterson, Eric Peterson, Bruce Ramsey, Daniel Rieke, Dave Ross, Joel Sacher, David Schuttenberg, Scoot!, Cindy Shattuck, Peter Shattuck, Terry Shattuck, Lois Shattuck, Dan Shattuck, Chris Shaunger, Scott Skidmore, Scott Smallwood, Steve Smith, Jani Smith, Mike Stirrat, Shelby Stirrat, Scott Stovall, Pam Strong, Jared Stuhlsatz, Bill Swinyard, Mark Vandament, Victor Voris, Kimmy Walker, Phil Watters, April Whitney, Gretchen Whitworth, Dana Wilson, Jaimie Wilson, Kristi Wolfer, and Lance E. Young.

And those I failed to mention who helped keep myself, my scoot, and my shop up and running all these years.

Contents

Foreword—Michael and Eric Dregni, 8
Introduction—Colin Shattuck, 11

Chapter 1 *The Evolution of a Revolution,* 13
Chapter 2 *Scooter Breeds,* 53
Chapter 3 *Can You See the Real Me,* 81
Chapter 4 *Runs, Rallies, Raids ... Mayhem,* 111

Appendix, 138
Index, 140
Bibliography and Art Credits, 142

Foreword

Let's get one thing straight from the get-go: If we have to explain the wonder and allure and coolness of motorscooters, you simply won't understand. Presumably, though, you've already seen the light, or you wouldn't bother flipping through this copy of *Scooters: Red Eyes, Whitewalls & Blue Smoke*.

I first saw the light while living in Italy in the mid-1980s during the Dark Ages of motorscootering. At the time, old scooters weren't vintage, they were junk—remembered by Italians with the same fondness as a good carpet bombing. Thus it was with great surprise to everyone when around the edge of the *Campo in Siena* came a primeval buzzing sound and into the piazza rode a two-tone, red-and-cream Lambretta Li150 Series II. It was like Earth suddenly halted its rotation, a hallelujah chorus of angels was cued, putti and cherubim flitted about, and a golden spotlight shined down to announce, "Behold ... cool."

Yet returning to the United States shortly thereafter, the Dark Ages were still in full bloom. Being the home team, Cushmans were the only thing anyone cared for or knew about. (And those Golfsters, man, they were nifty machines.)

"Lambretta? Didn't they build automobiles akin to Ferrari?"

"Parilla? Oh yeah, I've got Parilla tires on my car."

"Moto Rumi? Are you talking about that Islamic sex poet, boy?"

Happily, though, at that time scooters could be found anywhere and everywhere in barns, suburban garages, stashed in sheds. And they were cheap. When it comes to scoots, my brother, Eric, and I are Siamese twins (albeit born six years apart), and we bought our first scoots as a pair—a 1963 Lambretta Li125 Slimline and 1956 LD125—for $200. Total, that is.

Vespas came later, $50 for early Cruisaires being top dollar we'd pay. We casually perused weekend newspaper classifieds—*casually,* because there was no hurry as no one was going to beat you to a find—to pick and choose the sale scoots worth examining. Five hundred dollars for a batch of four running Vespas and attendant boxes of NOS spares? No way! I remember passing on a Vespa sidecar for $100: Hell, when you could buy a Vespa for $50, why pay double for a sidecar? I also recollect piloting home a Li150 Series II that started on the second kick after sitting for a couple decades in a garage and feeling queasy that someone had been ripped off after paying a dollar a cc for the beast. Even in Italy, a friend *gave* me his ancient TV175 Slimline; he had a brand spanking new Fiat automobile, so what'd he want that old scoot for?

Sadly, the market soon went through the roof. We were forced to cough up $150 for a 1957 Vespa—and the owner would only let it go if we carted away two Bulgarian motorcycles (one with a dead mouse in the crankcase), which I later pawned off on a buddy for a six-pack. We then suffered sticker-shock when a fledgling mod acquaintance paid an outrageous $300 for a Li125. The previous owner, Vern, was glad to pocket the money, scoffing at the pokey scoot, "If you want speed, get a Gold Wing!" The vintage putt-putt boom was being kickstarted into life ... .

Without really trying and despite the rising costs, my brother and I soon had a basement and garage full of the things. We happily proclaimed that the weekend didn't begin until the first scoot was kicked over, rarefied blue exhaust filled the Friday-night air, and the full-bore roar of a two-stroke alerted the neighbors that good times were about to start rolling. Our workshop was proudly named "Breath-of-Exhaust Lambretta De-Tuning Works," as everything we attempted to tune seemed to go slower afterward. To cure these ailments, we simply bolted on bigger carburetors, sawed off the muffler pipes, and blasted down backstreets avoiding the cops.

Motorscooters meant a gloriously misspent youth. Wrenching on and riding scoots was like learning a secret art as valuable as gunching a pinball machine or strumming garage power chords on a Stratocaster. (Again, if we have to explain, you won't understand.) I remember my brother's ongoing saga of a love life aboard his Lambretta. Eric had this gorgeous Greek girlfriend, and they'd go out for an evening only to find the scoot wouldn't actually scoot when midnight came—it had mysteriously changed back into Cinderella's pumpkin. He likely clocked hours trying to kick that thing over while his Greek girlfriend patiently waited on the curb, then walked home when it all proved futile. Yet when the Lambretta did work, it was like a vibrator on wheels.

By accident, somewhere between our youthful debaucheries tinkering and riding and our ceaseless hunt for scooters, Eric and I wrote our first book, *Illustrated Motorscooter Buyer's Guide*, published way back in 1993. See, many scooters we bought came complete with original brochures, photos, and other memorabilia—anything the old owner could dump on us. We owe you all an apology for that book. The thing was, at that time there was little to no information out there, and researching consisted of rumors, wild-goose chases, and gutting brochures for bore and stroke stats. We made our share of errors—both of fact and omission—but we hope it at least provided inspiration to the handful of putt-puttniks around the globe.

Our proudest moment in scooterdom, though, was the scooter club coup that came around the time of our book's publication. We here in St. Paul, Minnesota, had a loose-knit scooter club that I guess my brother and I started without planning to. That club, though, was all about to unravel. The few scooterists in town showed up with ancient parts lists and decrepit owners manuals; they wanted to talk shop while we just wanted to drink coffee and ride. We named the club Boltstrippers

As the result of a misspent youth, Michael and Eric Dregni are the proud authors of three histories of motorscooters—including The Scooter Bible, Scooters!, *and* Illustrated Motorscooter Buyer's Guide*—as well as other esoteric books on a variety of eccentric topics. Other scooteristi may scratch their scalps trying to figure out the strange sense of humor running through these books ,like two-stroke miscèla through a Lambretta engine on the verge of seizing, but the brothers Dregni are always happy to examine the quaint and curious sides of scootering before bothering to check the oil. Native sons of Minnesota, they can fake a good Milanese accent and roll their spaghetti with the best of them: Michael lived for a time in Siena while Eric spent several years in Brescia and Modena, all providing fine fodder for their scooter histories. They currently live in Minneapolis (within two blocks of each other), both are married to beautiful brunettes (with the same haircut), and both have wild, out-of-control children dying to get their hands on the keys to the Vespas.*

Anonymous after our oft-misguided handiwork torquing nuts beyond the breaking point. Instead of analyzing scooters, my brother was more interested in writing an operatic biography of Albania's King Zog and penning fan letters to Winona Ryder (to the fury of his aforementioned Greek girlfriend—a *fury* he should have been paying more attention to, viz. Euripides et al.) while I simply wasn't interested in Vespa-nerd brake-swept-area comparisons. Obviously for the new scooter crew, we weren't *serious*. Next thing we knew, we were cast out of the scooter club we ourselves formed and they began secretly meeting elsewhere on a different night. Already, we were branded scooter heretics.

For us, scooters have always been about fun as well as the mechanicals and history and coolness. Colin's book brings that fun to life. He tells the story of both scooters and the scooter faithful—all of us scootering brothers and sisters. Read his book and enjoy: Two-strokes have been resurrected, wrenches ratcheted, clubs united, rides ridden. The golden days of scootering are back!

—Michael and Eric Dregni

The Denver crew who started Colin Shattucks's scooter journey

Introduction

I can remember being ten, riding along side my father in the front seat of his 1965 Buick Electra 225. It was a massive car. It had four doors, each the size of an average cattle-barn door. Rescued from within the barbed-wire confine of the local police auction one sunny morning, it was beat to shit. To look at the thing you wouldn't have suspected that it was roadworthy, but it was, barely. It consumed oil almost as quickly as it did gasoline, but its 401-cubic-inch v-8 still chugged down the highway with inefficient ease. Dad had it equipped with a huge front bumper built from steel tubing and welded directly to the chassis so that the beast could be used to literally push other vehicles around. We called it "the bumper car." My father once shoved a city bus out of a snow bank with it. Seriously.

My dad was a stellar example of how Americans love large, very shiny things. Especially cars. Oh, there was a while there, a decade or so, when, out of sheer necessity, we adjusted the size and efficiency of our vehicles to conserve resources. Cars got smaller and lamer and smaller and goofier and smaller and more disposable. From the time I was about eight to the time I got out of high school, cars, new cars, American cars, sucked. *Hint: I was born in 1972.*

For those of use raised in the seventies, brand new cars were simply a reminder of a dull and grim reality. Times were tough and cars were tools. We did not anticipate each year's new models. We were the "$88-bucks-a-month,-push,-pull,-or-drag,-no-money-down-but-the-car-is-a-piece-of-shit" generation. This left me looking for more. Old cars were my dad's thing; motorcycles were too "cool."

I was a different kid, strange. I would walk the endless route to junior high school alone, imagining a little personal vehicle that was sort of like a spaceship—shaped like a tiny Tylenol capsule, but with wheels—that I could travel around on and show off to my friends. I could paint it up really cool and it could be my freedom machine. I wanted a car with two wheels, a *little* shiny car.

Interestingly, it was during this same era, that which brought us disco music, gas lines, and bell-bottomed jeans, that something else occurred right under most of our noses. It seems that those little vehicles I had fantasized about had already existed for decades and that America was catching on. People, young and old, from coast to coast, were discovering a fun not-so-new form of transportation, one that was economical, practical, and fuel efficient all the while providing them with an escape. They were buying them by the thousands. Americans had discovered what Europeans already treasured and I had never heard of … *scooters.*

—Colin Shattuck

Chapter 1

The Evolution of a Revolution

In the beginning, a scooter was nothing more than a child's toy with an engine affixed. Uncomplicated, somewhat dangerous, and quite often unreliable, the first scooters bore a greater resemblance to prehistoric Go-Ped prototypes than to the full-bodied, sit-down models that came later. Scooters managed to carve their niche in the marketplace by delivering in times of hardship. Even though they owe much of their lineage to war, economic depression, and urban overcrowding, scooters still slipped into the hands of those simply looking for a joy ride.

An idea had been hatched for a type of vehicle that was not a motorcycle or an automobile or a bicycle, but one that combined the best assets of each. Scooters offered the open-aired freedom and adventure of a motorcycle, the lightweight ease and efficient simplicity of a bicycle, and the skinned, polished, and painted veneer of an automobile, all in an affordable package. As the years progressed and technology advanced, scooters evolved from simply practical and economical to beautiful and powerful. Never lost in this evolution was the scooter's key attraction—fun—even though the circumstances that led to sales booms were not always such.

The First Scooter Craze

The American Motoped, first produced in 1915, was a sign of motorscooters to come. The Motoped, unlike the motorcycle-like French Monet-Goyon that actually predated it, was based around the very same concept as kids' push scooters, only a small displacement engine over the front wheel replaced the pushing action of the foot. The next popular American scooter was the Autoped. It was built in New York beginning in 1917 and gained international recognition for its potential as a lightweight military vehicle that could be dropped with paratroopers from an aircraft. However, these early machines were crude and ultimately failed to make a major impact.

Across the pond, the British firm ABC created its Scootamota in 1919. Certainly a step beyond the Autoped and the Motoped, it was much larger and more useful. While the Scootamota retained the basic skateboard-like floor of the early American scooters, it featured a motorcycle-like design, with larger wheels, a much bigger and more powerful engine, and a top speed of twenty miles per hour. The first scooter with glamour appeal, the Scootamota posed for photographs with some of the

(opposite) Two boys on Burl Ives' scooter, New York, 1955

biggest British celebrities of the day. Its design prompted imitators in England as well as in Germany and France. It continued production until 1923, when scooters essentially disappeared from Europe and America for the next decade. The early scooter builders developed the basic framework for the scooter that would eventually become a key part of a successful formula, just not for another decade or so. America had introduced an entirely new type of vehicle, one that was just a couple of decades ahead of its time.

Cheaper Than Shoe Leather: American Scooters of the Great Depression

An American scooter is, by definition, a vintage scooter; the last American-built scooters with any real market impact date from the early 1960s. During the 1930s, American-made scooters entered the mainstream as they became reliable forms of alternative transportation. They would eventually be outclassed and outpriced by European and Asian models and fade away to obscurity. In their heyday, though, American scooters were the best in the world, and huge numbers of people still adore them.

The likelihood that one might encounter one of these people increases proportionately to one's proximity to Lincoln, Nebraska. The home of the Cornhuskers was once the proud host to a corn-fed, pipe-bending, motor-making factory that, in the 1930s, began a three-decade legacy as the greatest American motorscooter maker. Cushman Motor Works entered the scooter fray out of necessity, the father of invention in this case being the Great Depression. The Lincoln plant sold mostly engines at the time, including Cushman Husky engines for power equipment applications. Demand fell sharply during the notorious lean years and Cushman was receptive to the idea of selling thousands of Huskies to a man by the name of E. Foster Salsbury. Salsbury was seeking power plants for his newly designed scooter, the Motor Glide. He sent blueprints of his design to Cushman seeking to purchase a thousand Husky engines. Whatever Cushman's price, Salsbury did not wish to pay it and he decided to instead use an Evinrude engine in the Motor Glide. Production began with the Evinrude, and Salsbury quickly began promoting his scooter, seeking people of fame to tout its benefits.

According to *Illustrated Motorscooter Buyer's Guide*, Colonel Roscoe Turner, a famed barnstormer, became an early Motor Glide endorser. When he appeared in Lincoln, just miles from the Cushman plant, to perform his aerial show he rode a Salsbury Motor Glide around the airfield. Lincolnites took notice. One local boy was so inspired by the sight that he rushed right home to a scooter of his own, which he fashioned from scrap and powered with a Cushman washing machine engine.

Apparently, the boy appeared numerous times at Cushman Motor Works buying parts for the contraption. One day, Cushman honcho Charlie Ammon noticed the boy and his scooter through the window of his office. He was impressed by the boy's inventiveness and his mind started whirring. Robert Ammon, his nineteen-year-old son, helped convince the conservative Charlie that the idea of building scooters would be extremely good for the struggling company.

It was not difficult for them to design scooters. They were relatively simple machines constructed primarily from parts the company was already producing. And they had the blueprints that Mr. Salsbury had so generously provided. In a 1984 tribute article in the *Lincoln Journal Star,* Robert Ammon said that they had the prototype Auto-Glide working within thirty days of starting the project. "It was a very crude-looking thing, of course," he added.

Cushman and Salsbury produced scooters in fierce and direct competition between 1936 and 1940, under the respective model names Motor Glide and Auto-Glide. While Salsbury relied on celebrity endorsements, Cushman advertised the technical superiority of their Auto-Glide, calling it, "the greatest advance ever made in low-cost, motor transportation." The economic benefits were front-and-center in Cushman's advertisements as well. The machines were touted as being "cheaper than shoe leather" and operated at "one-tenth the cost of an auto." Cushman promised thirty miles per hour at 120 miles per gallon, numbers that are still admirable today. Salsbury touted economy as well as boasting that the "Moto Glide will take you places at the, lowest cost."

From the start, Cushman had an advantage. Their Husky engine was quite good, perhaps the best engine available for a scooter. After all, it was the Husky that led Salsbury to contact Cushman in the first place; he knew that Cushman was producing a superior power plant known for its reliability, quality, and power. Huskies had been on the market for well over a decade before they were used to power scooters. Early Auto-Glides produced 1.5 horsepower, twice that of a Motor Glide. Auto-Glides also featured a superior chain drive where the first Motor Glides utilized a flawed roller that failed when it got wet.

A fortunate traffic officer ticketing speeding scooterists, Oakland, California, 1937

By 1938, however, Salsbury made very dramatic improvements to the Motor Glide that helped stoke the rivalry between the factories. They switched to a 1.5-horsepower Johnson engine with a chain drive. They also added the feature that would make the Salsbury a legend: the automatic transmission. Salsbury referred to this advancement as the, "self-shifting torque converter." The basic design of this centrifugal transmission was well ahead of its time in 1938. Known today as a CVT, or "constantly variable transmission," the belt-driven, pulley-operated automatic drive has appeared in a wide range of machines over the past eighty years, including lawnmowers, ATVs, and automobiles. And the CVT's legacy lives on. Almost every new scooter built today uses Salsbury's basic design.

Not to be outdone by their rivals, Cushman debuted an automatic of their own in 1940. Cushman referred to it as the "floating drive" and claimed that it was, "just like the fluid drive on the latest models of automobiles." It wasn't. It was centrifugal, just like Salsbury's, only different in design.

Cushman and Salsbury exchanged technological and promotional blows during the Great Depression and through World War II with moderate competition from other manufacturers. New scooters were popping up like dandelions, with names like Rockola, Zip Scoot, Auto scoot, Trotwood, and Constructa-scoot. The most formidable was Chicago's Moto-Scoot. Founded in 1936, the small three-man operation constructed scooters that closely resembled those of Salsbury and Cushman. Moto-Scoot, like many competitors, advertised in the back of *Popular Mechanics*. The ads worked and the company grew quickly. By 1939, Moto-Scoot had a new factory and twenty-five workers. Sales were brisk and, in 1939, *Time* proclaimed the company's founder, Norman Siegal, to be the "Henry Ford of the Scooter Business." Accolades aside, Moto-Scoot would become a footnote in scootering history, largely because of the company's shift in focus to other types of manufacturing during World War II.

Despite the small number of surviving makers, the late 1930s proved to be the Golden Era of the American motorscooter. The technological advancements that were made then proved invaluable to the industry worldwide and for generations to come.

World War II:
Scooters Rain from the Skies

World War II proved a huge boon to the American scooter industry. Automakers like Ford were dedicated exclusively to building vehicles for the military from 1942 to 1945, and Harley-Davidson and Indian were furiously manufacturing motorcycles for the war effort. This left scooter builders like Cushman and Salsbury a huge underserved market, American civilians. Both companies received exemptions from the war department to offer their products to the general public. With fuel, steel, and tire rubber extremely scarce, riding a scooter became a very patriotic thing to do in the early 1940s.

From 1942 to 1945, Cushman produced the 30 Series models, the last to be referred to as Auto-Glides. (After the war, the company would brand their machines as Cushmans in the interest of promoting the company name.) Available as a manual or an automatic, the 30 Series saw the engine output upped to four horsepower. These machines were very popular on the home front and sold well.

Salsbury also sold large numbers of scooters during that period. Their Model 72 was streamlined and luxurious compared to earlier efforts and its available range of accessories was unrivaled. A buyer could choose from accessory seats, carriers, sidecars, windshields, towing packages, electric horns, and custom paint colors. Gas rationing helped spurn stateside sales of both the 30 Series and the Model 72; both were commonly seen in the cities and in rural America.

Both companies also made their mark on the American war machine, though Cushman to a much larger degree. While Salsbury sold a handful of Model 72s to the U. S. Navy for base transportation during the war and also offered an ambulance version of the Motor Glide, Cushman shifted the majority of its production to military use. They produced two- and three-wheel scooters for the Army, Navy, and Army Air Forces.

In 1943, the American military sought bids to produce scooters for use as airborne vehicles to be dropped with parachutes from aircraft so that landing troops would be equipped with ready ground transportation. This tactic was already in use by the Italians and the British. Italy had the Volugrafo paratrooper scooter and the Brits had the Welbike, a compact folding scooter ideal for the chore. Cushman provided prototypes of their newly named Model 53 and Cooper Motors of

Albert G. Crocker's Scootabout added beautiful lines to the scooter, 1939

Los Angeles offered the Cooper Combat Motor Scooter. Cooper's factory was not up-to-snuff, however, and the military put its faith in the more proven Lincoln plant of Cushman Motor Works. By the end of 1945, Cushman had sold more than 4,000 Model 53s to Uncle Sam. The six-inch tires fitted to the 53 were interchangeable with Air Force spotter planes and the machine was made available in both olive drab green and bright yellow, to

Commuting to work the futuristic way, on a Salsbury Model 85 Superscooter

be unseen and to be easily seen, depending on their use. Model 53s were simple in design and featured no lighting or suspension, luxuries not deemed necessary on the workhorse scooters. They were involved in battles in the Mediterranean and Pacific theaters and were dropped on the beaches of Normandy.

Cushman was only one of many world manufacturers building scooters for military use during the war. Many European manufacturers also provided their nations with scooters to meet all kinds of military transit needs including those of paratroopers. The war itself would not, however, be as great an impetus to the scooter industry as its eventual end would be. The needs of civilians in the wake of the war, combined with the destruction of the European infrastructure and the excess production capacities left dormant, resulted in the real scooter boom.

Business as Usual for Postwar U. S.

Conditions after World War II were far different in the U. S. than they were in other scooter-producing countries. America and its allies had won, its citizens were ecstatic, its infrastructure was unscathed, and its economy was humming. Meanwhile, European nations lay in ruins. Cities and infrastructures were badly damaged, factories bombed, roads and bridges destroyed. Their economies were blighted and, save for America's reconstruction efforts, hopeless.

Back home, Cushman and Salsbury thrived. Their production lines, freed from their wartime duties, cranked out scooters for the masses. Cushman did not need to retool their factories after the war. They offered the public the same basic machine they offered the government, only with prettier paint, shinier bits, and more accessories. They dubbed their once-militaristic machine the "family scooter." According to Robert Ammon, sales of Cushman scooters, "really took off, right after World War II."

Salsbury was not content to simply re-badge their war-era scooters, and instead got radical and futuristic with their Model 85. It was the perfect design for a nation looking to the future, its appearance straight out of a Buck Rogers comic, streamlined and smooth. It was also gigantic, closer in size to a small car than a small scooter. The rider sat entirely shrouded behind the bubble-like front; its long, swooping tail extended far behind the seat. It was low, slow, and covered in paint; a little, shiny car at a time when most people could still not yet afford new cars. Following Cushman's lead, Salsbury dropped the name Moto Glide and opted to promote the Salsbury brand.

For a few years after the war, Americans continued buying scooters in great numbers. Cushman and Salsbury, along with a host of less significant companies, jostled for the hearts and minds of U. S. consumers. In this early salvo, Cushman outlasted Salsbury and was declared the winner. Neither company, however, would go on to take the bigger war. In fact, the American market became less significant as scootering entered its postwar era. Instead of little, shiny cars, the masses were ready for great big ones. Families could afford automobiles again, and Ford, GM, and Chrysler prepared to bring Americans into the era of the hood ornament, the porthole, and the tailfin.

The Scooter Renaissance

In postwar Italy, the sight of American soldiers tooling around cities and Army bases on reliable little scooters was exciting and inspiring. One Italian firm had attempted to build scooters for the Italian military during the war, but never reached viability. But the new Cushmans proved very useful and the soldiers seemed to enjoy riding them, too. Among the Italians who took notice were Ferdinando Innocenti and Enrico Piaggio, the industrialists who would lead all of Europe into the motorscooter renaissance after World War II.

By building the bullets, bombs, and aircraft, both Innocenti and Piaggio made their companies crucial targets of the Allies and saw their factories all but demolished. But these gutsy entrepreneurs refused to allow their respective companies to perish. They shared a vision: scooters were ideally suited for postwar Italy, and they realized that scooters could revive their companies in the wake of World War II. Both firms could utilize technology that they already possessed. Piaggio and Innocenti quickly rebuilt and retooled, and then set about the task of reinventing the scooter, retaining its practicality and simplicity, while giving it flair that was distinctly Italian.

Italy became the motherland of the motorscooter between 1946 and 1948 and to this day has never relinquished claim to the title. The story of the postwar

Italian racers on Vespas, 1953

Italian scooter renaissance is a tale of two companies. Much like Cushman and Salsbury jostled during the Great Depression, Piaggio and Innocenti would spend the years following the war battling for each precious chunk of market share. Much has been written about Piaggio and Innocenti and the legendary scooters they created: the Vespa and the Lambretta. It is one of the great corporate contests in manufacturing history, and the scooters borne of it are the most treasured jewels in all of scootering.

Piaggio

Outside of Pisa on on the banks of the Arno River, sits a quiet factory town named Pontedera. It is the home of Piaggio, birthplace of the Vespa. Road signs throughout the town simply reading "Vespa" point visitors toward a massive factory. Resembling a military compound and surrounded by stone walls more than ten feet high, the Piaggio factory comprises a large section of Pontedera. Giant smoke stacks climb from the factory roof high up into the Tuscan sky. Each features a large, blue six-sided shield containing the famous Piaggio logo, two stylized letter Ps, back-to-back so that they appear to be mirror images of each other. Together, the two Ps resemble an airborne wasp ready to strike; appropriate because the word *vespa* means "wasp."

By the time the Vespa was born, its parent company had already been in the transportation business for more than fifty years. Founded in 1884 in Genoa by then twenty-year-old Rinaldo Piaggio, it was first called Societa Anonima Piaggio. They began by manufacturing woodworking tools, then fittings for ocean liners, and later truck bodies, railcars, and train engines. World War I further diversified Piaggio's business into aircraft. The firm is credited with inventing pressurized cabins for airplanes as well as retractable landing gear. Piaggio's national importance grew as Italy entered World War II. The company designed and built the legendary P-108, fascist Italy's only bomber, at its Pontedera facility.

It was Piaggio's military importance that made it successful during World War II. It also made the company and its facilities targets of great importance to Allied bombers. In August 1943, bombs leveled the company's plants. Enrico and Amondo Piaggio, sons of the company's founder, were left with the challenge of rebuilding their once strong company. Enrico was left responsible for rebuilding the ravaged Tuscan factories in Pisa and Pontedera. In 1957, Enrico told *American Mercury* magazine, "Our over 10,000 employees were thrown out of work by the bombings and by the fact that, as soon as the war was over, our production fell to zero. In fact, we were prohibited from making airplanes by the peace treaty. So you see it was essential that we find a new peacetime product for the sake of the Piaggio Company and our employees."

Inspired by the needs of his company, his employees, and the Italian people, Enrico decided that the future of Piaggio was clear and simple: the company would manufacture lightweight, cost-effective, convenient … scooters. He enlisted a trusted aeronautical engineer by the name of Corradino D'Ascanio, the man who first built a working helicopter. Enrico asked him to design and build the first Piaggio scooter in 1946, based on the machines first seen falling from the sky during the war. Nicknamed "Paperino," the Italian name for Donald Duck, D'Ascanio's design did not please Piaggio. He found it quite distasteful, an ugly little duckling as it were. The idea was right, though, and Enrico asked D'Ascanio to try again.

D'Ascanio made it known that he hated motorcycles. He found them bulky, dirty, and unreliable. He wanted his new design to more closely resemble an automobile, and his aeronautical experience provided him with the tools to create a completely new product. At the heart of his unique design was the "monocoque," a single-piece, pressed-steel body unlike any two-wheeled chassis ever made. He designed the seating position to be low, allowing riders of all sizes to feel comfortable, and gave the scooter a rear-mounted 98cc engine that would not expose the rider to excessive heat. He also eliminated the drive chain, which he felt was a filthy flaw in traditional motorcycles. He located the gear changer on the handlebar, a forward-thinking element that would immediately become the standard for scooters. D'Ascanio presented drawings of this new design only a few days after the Paperino was rejected, and Piaggio told him to build it. Production of a prototype began immediately, and in April 1946 the MP6 was ready. Piaggio looked at the prototype and happily exclaimed: "It looks like a wasp!" The Vespa had been born.

Innocenti

The Innocenti family was in the business of blacksmithing. Its patriarch, Dante Innocenti, was well known in the small Tuscan town of Pescia. He provided hardware for the locals and became a successful businessman. Dante's son Ferdinando was cut from the same cloth. The blacksmith's son grew up to be a master pipe-bender and revolutionary industrialist. Emboldened by the fascist government's directives to jump-start Italian manufacturing, Ferdinando opened a factory in the late 1920s.

The company was known as Fratelli Innocenti, or Innocenti Brothers. In 1931, the upstart firm signed a contract with the Vatican to design a series of drainpipes for the Papal Gardens of Castelgandolfo. In 1933 the firm would sign a lucrative agreement to provide steel scaffolds for the construction of large buildings. The revenues from this deal allowed for the company's rapid expansion, and Ferdinando decided to build a new factory in Milan to complement its plant in Rome.

By 1935, in anticipation of what seemed an unavoidable war, Innocenti upped its staff at the Milan plant from 100 to 500 workers. The company shifted its focus to aircraft hangars and bomb casings. Like Piaggio, Innocenti's production would become exclusively dedicated to the war effort during World War II. Rather than building aircraft and engines, Innocenti built smaller munitions. A new plant built in Rome in 1940 produced 40,000 bullets per day. Innocenti's efficiency earned the company the distinction of being called a "Model Fascist Factory" by the Italian government.

In April 1944, bomb after bomb fell on Innocenti's factories, bringing the company to its knees. It would produce nothing until 1946 when the massive rebuild of their plant was undertaken.

In 1944, when Rome was liberated from the Nazis, the idea entered Ferdinando Innocenti's mind to build a scooter. He saw the American troops getting places on their Cushmans and realized that similar scooters could be built using parts already in his factories. He set out to design a scooter that very year, two full years *before* Piaggio.

Ferdinando sought to have a design ready before the war ended so that he could begin production immediately. He too contacted the well-known designer Corradino D'Ascanio about the concept of designing a scooter like the ones he had seen in Rome piloted by Allied forces. He and D'Ascanio agreed that Innocenti should get into the business of building the lightweight people movers as soon as possible. But D'Ascanio never actually designed a scooter for Innocenti; instead he would move on to assist Piaggio with the Vespa in 1948.

Ferdinando remained inspired by the notion of creating his own scooters. Undaunted, he hired a Roman designer to begin work on a scooter roughly based on the Cushman Model 32. The designer used the shape of a torpedo as the basis for the styling of the machine, instead of the chunky and utilitarian look of the Cushman. "Experiment O" was the name given to the original prototype design. A model was completed quickly and

A loaded transport leaves the Lambretta plant piled high with three layers of motorscooters, 1955

unveiled in the form of carved wood. This early design featured an aeronautic pressed-steel body, later the easily identified trademark of the Vespa. But the "Experiment O" scooter was never built, possibly because Innocenti was not pleased with the design, but probably because of the chaos that continued to rage on.

In 1945, Innocenti turned the scooter initiative over to another successful aeronautical engineer by the name of Pierluigi Torre who started from scratch. Torre's scooter featured a rigid tubular backbone of iron, off of which other components hung. The engine was placed in the rear of the scooter, underneath the seat near the fuel tank and glove box. While work on the prototype continued, Innocenti was finally able to return to his Milan factory and see the devastation. Besides the plant being rendered virtually useless, Allied forces had

The Rumi Formocino was among the fastest (and strangest) Italian scooters

commandeered it. While Innocenti wrestled with the Allies for control of his plant, Piaggio began producing Vespas. In 1946, Ferdinando finally regained control of his factories. He had lost ground in the scooter wars, but he was not about to give up. Innocenti envisioned using the American "assembly line" concept to catch up. His factories would be the first in Europe to incorporate this ground-breaking method of manufacturing.

With his factories back in his own hands and being rebuilt, Innocenti was ready for the final design of his scooter. After months of nearly round-the-clock design work, a prototype was revealed to Ferdinando, who felt it needed only minor cosmetic revisions. Rather than conceal the mechanics with steel like Piaggio, Innocenti decided to make the mechanics stylish and incorporate them as key design elements. "Experiment 2," as the revised machine was known, was first built in 1947.

Innocenti had a scooter, now he needed a name. According to the official Innocenti newsletter, "Lambretta is the nickname of a small creek that runs through the Innocenti factory." In honor of the waterway, Innocenti dubbed his scooter the "Lambretta M"—the "M" for motorscooter. The Vespa had its rival.

Vespa and Lambretta Take It to the Street

Innocenti had beaten Piaggio in realizing that he should build scooters, but Piaggio won the race to market. It was the beginning of a period in which the two companies would stop at nothing to beat one another.

Automobiles were very scarce in Italy after the war, as was gasoline. Scooters were seen as affordable and economical, not to mention stylish. The public quickly began to embrace them, in particular, women. Enrico Piaggio had decided the Vespa should feature a step-through design so gals could ride them while wearing a dress. The motorcycles of the era were heavy, tall, and not dainty in the least. The new Vespas and Lambrettas were significantly friendlier, cleaner, and less intimidating.

Scooters appeared as though they were the right vehicles at the right time. The very low cost of the machines was perhaps the biggest reason for their instant popularity. In a 1956 interview in *Newsweek*, Enrico Piaggio emphasized the importance of the Vespa's affordability, touting, "We put automotive transport within the reach of people who never expected to travel that way."

Vespas were available to the public by April 1946. The earliest models, known today as the Vespa 98 or the rod models, were branded simply "Vespa" with no specific model designation. They were design marvels, as beautiful as they were radical. They featured a left-hand twist-grip shifter connected to the rear-mounted engine by a series of steel linkage rods. Piaggio raved that the Vespa could reach thirty-five miles per hour while only sipping gasoline at a rate of 100 miles per gallon.

Soon Innocenti began a radio campaign declaring that it was "Lambretta Time," despite the fact that it would be many months before the firm was actually ready to sell any scooters. Innocenti was simply attempting to give consumers pause, encouraging them to wait for their new machine rather than immediately heading down to the nearest Vespa dealer. Lambretta time would finally come in October 1947, one year and one summer after the Vespa. No longer called the "M" model, the first Lambretta was known as the "A" model. Innocenti would build more than 10,000 the very first year of production. In contrast, Vespa produced more than 30,000 units by the end of 1948. Unlike the Vespa, the A had its shifter mounted to the floorboard. It was a rocking heel-to-toe style mechanism. A marker on the floor told the rider which of its three gears they were in. Innocenti equipped the A with a 125cc engine that was slightly larger than Vespa's 98cc mill. These would be the first in a wide string of engine displacements offered by the two companies over several decades.

Innocenti, it seemed, always pushed the engine sizes upward in an attempt to offer consumers more power and greater top speeds. The A produced 4.3 horsepower to the Vespa's 3.3 horsepower. It also carried a passenger with the addition of an optional pillion pad that could be attached to the top of the tool case on the rear. This was the Lambretta's most effective selling feature. But the A had smaller, less stable wheels than the Vespa, however, and this became a major consumer complaint. It also lacked rear suspension.

There were other contrasts. Early Lambrettas were very simple looking and machine-like, while the Vespa was sculpted and elegant, streamlined and sexy. Innocenti elected to make the engine and the other mechanical bits prominent cosmetic features on their scooters while Piaggio shrouded them almost entirely with steel.

In 1948, Innocenti replaced the A model with, you guessed it, the "B." In order to keep up with Piaggio, Innocenti adopted the left-hand twist shifter, bumped the wheel diameter from seven to eight inches, and added a fancy new rear shock. The B would remain Innocenti's state-of-the-art scooter through 1950. Innocenti was building 100 scooters on a good day.

The ever-competitive Piaggio replaced their 98cc engine after two years of faithful service. Naturally, they bumped the displacement to 125cc. This new engine offered 5 full horsepower, rendering the Vespa mightier than the 4.3-horsepower Lambretta B. Top speed now exceeded forty miles per hour. In 1949, Vespa replaced their shift linkage rods with control cables similar to those still used today.

The scooter race had truly begun. Though neither Innocenti nor Piaggio would make any radical changes for a couple of more years, they had each begun to establish separate followings, each of which was plenty passionate in their scooter preference. Over the next several years the scooter club phenomenon began and scooter races and skill competitions spread across Italy. The nation collectively began associating scooters with better times. Piaggio and Innocenti had started something truly special, something that was re-sculpting Italy's identity and moving its people in more ways than one.

Moto Guzzi Galetto

Rumi Formichino

Motobi Catria

Aermacchi

Reconstruction Means Scooter Construction: Postwar Scooters of Europe and Asia

The brand name "Vespa" has become virtually synonymous with the word "scooter." It is the first word to enter one's mind when scooters are mentioned. Despite its familiarity, however, a Vespa has never been the only scooter to choose. Every great idea has imitators and every growing market will draw competition, and, in the case of postwar scooters, the competition was healthy. As the dust cleared after World War II, many companies throughout Europe and Asia were inspired by the famous "wasp" to design and build scooters of their own—some great, some not-so-great.

In Italy, several firms followed very closely in the footsteps of Piaggio and Innocenti. Plants previously dedicated to building things like refrigerators and sewing machines became frenzied scooter factories, clamoring to build as quickly as consumers were clamoring to buy. The trend spread across Europe. Germany became a hotbed of scooter manufacturing along with Britain, France, Spain, Czechoslovakia, and even Russia.

Interestingly, as one scooter revolution was taking place in Europe, its Asian counterpart was developing in a near mirror image, spawned by the familiar needs of reconstruction, putting people to work, and getting from point "A" to point "B." Two Japanese firms—Fuji and Mitsubishi—gave birth to the Asian scooter industry in precisely the same way Piaggio and Innocenti had the Italian, and did so almost simultaneously. Separately, basically unaware of each other, the European and Asian scooter markets blossomed after World War II, and the speed with which the products improved was stunning. The world was gaining options, choices.

Other Postwar Italian Scooters

Lambretta and Vespa are far from the only names worth mentioning in a history of these addictive little machines.

Perhaps the twin titans' most elegant Italian competitor was ISO, a company noted for manufacturing refrigerators that went into the scooter trade from 1948 to 1960. Parilla, while better known for motorcycles, is another Italian company that worked the postwar scooter market. Parilla's scooters were never as successful as the German models they inspired, Zundapp's Bella models, but they did sell relatively well. Other Italian scoot brands include the short-lived Moto Rumi, whose speedy Formocino, or "little ant,"is one of the most unique and bizarre scooters ever conceived, and MV Agusta.

German Scooters

Not surprisingly, German engineers developed some of the most refined and sturdy scooters ever built. The Walba scooter was among the first, hitting the market in 1949. Only a few brand names from Germany had a substantial impact, including NSU, which manufactured Lambrettas under license that outdid their Italian-made counterparts, especially the company's masterpiece, the Prima Funfstern; Heinkel, the airplane maker that dabbled in scoots, including the sturdy Tourist; and Zundapp, whose Bella drew its inspiration from Parilla, not Vespa or Lambretta.

British Scooters

Despite the islands' importance to the history of scooter culture, scooters built in the U. K. are essentially historical footnotes—and strange ones at that. The cigar-shaped Piatti is in a class by itself in terms of being just plain weird; scoots by Triumph/BSA, while not as odd, never enjoyed the popularity of their European peers.

Japanese Scooters

After the end of WWII, on the opposite side of the world from Italy, a very similar rivalry to that of Piaggio and Innocenti was born. Picking up the pieces and moving onto the road to recovery, Fuji and Mitsubishi entered the scooter market with a bang. Japan embraced the scooter as an alternative to the automobile in precisely the same way Italy had. By 1954 there were nearly a half million scooters darting through Japan. The two early Japanese

Heinkel Tourists remain favored by enthusiasts

A woman lifts up the seat to give access to the petrol tank in one of the first British scooters, the Dayton Albatross produced by the Dayton Cycle Company, 1954

scooter companies paved the way for two others, which would make their impact a couple of decades later. Those companies were Honda and Yamaha.

America Rediscovers Scooters

The 1950s were America's moments in the sun. Together with their Allies, the U. S. had won the war and the nation turned its attention toward rewarding itself, entertaining itself, and buying itself shiny new presents. The big four automakers were, once again, allowed to paint cars red, yellow, pink, and seafoam, not just olive green. The servicemen were home making babies, homes were growing, cars were huge, and folks were watching TV. Everything was coming up roses. Naturally, a country so drenched in success had little need for a vehicle that was growing popular, due to its practicality and economy, in the redeveloping Axis nations. Americans were not so concerned with pinching pennies at the gas pump and, with the exception of Pearl Harbor, had suffered no damage to infrastructure. The standard scooter-buying motivations were just not there. In fact, America's highway system was growing like a fat snake across the countryside.

For scooters to catch on in America, they had to be fun. Then, in 1953, an indelible image graced movie screens from coast to coast, an image that instantly transformed scooters from tools to toys in the minds of Americans: Audrey Hepburn, looking gorgeous, learning to ride a Vespa.

Roman Holiday, starring Hepburn alongside Gregory Peck, premiered in 1953 and won a phenomenal ten Academy Awards. Hepburn took Best Actress for her portrayal of a runaway princess adventuring through Rome, who meets a Vespa-riding hunk of an American named Joe Bradley (the fabulously handsome Peck). In one of the films most romantic scenes, Hepburn throws her arms around Peck's shoulders and rides on the back of his 125cc "handlebar" Vespa past

Rome's famous sites. Soon she takes the controls and weaves dangerously through traffic until the *polizia* stops them. Peck wins the couple's freedom by claiming that they were, "going to church to get married on a scooter." *Roman Holiday* was a runaway hit and Americans, particularly women, were enamored by its best supporting actor, the Vespa scooter.

Shriners have a long history of riding scooters in parades and exhibitions

An Eagle Spreads Its Wings

As America's soldiers returned home at the end of the 1940s, the Cushman/Salsbury era was still in full swing. Bolstered by their sales to the U. S. military, Cushman was particularly well positioned for a sales boom. Cushman's sturdy Model 53, a hit with the paratroopers and a catalyst for Piaggio and Innocenti, had returned home and was dressed in civilian clothing.

Cushman introduced its new 60 Series scooters in 1949. As the new scooters in Italy had been influenced by Cushmans, Cushmans were now influenced by Vespas and Lambrettas. The 60 Series scooters were sleeker than previous Cushmans and featured steel bodywork to hide the mechanical goings-on. According to *Illustrated Motorscooter Buyer's Guide*, "The bodywork of the clothed 60 Series was much like that of the 50 Series except it was smoothed over and rounded off. The rear was wider with a turtleback look." American scooters were beginning to look more like Italian ones, at least they were trying to.

Models like the Cushman Pacemaker and RoadKing were European-style, step-through scooters while the 60 Series retained essentially the same look as depression-era Cushmans. Each type sold well. Cushman reportedly built 10,000 scooters in 1950. It was when the Eagle landed that sales really picked up. "Someone in our sales department wanted a scooter that looked like a motorcycle with a gas tank between your legs," said Ammon. "It turned out to be a hell of an idea."

The Eagle Series was America's postwar scooter. Unveiled in 1949, the Eagle sported the appearance of America's big V-twin motorcycles like Harley-Davidson and Indian. Eagle models were a departure from the refinement of the motorscooter. No more were Cushman's engineers concerned with mimicking the Italians. The Husky engine was exposed in its full glory and the frame design required the rider to step over, not through the motorcycle-style tank that proudly bore a Cushman badge along with the word "Eagle" and a picture of a swooping bird of prey. The Super Eagle and the Silver Eagle would soon join the Eagle and the cycle-like scooters became the best selling Cushman series of all time. Charles Ammon bragged, in 1950, that his company owned 90 percent of the American scooter market. Cushman Eagles were produced all the way up until 1961 when Ammon's company finally succumbed to the deadly sting of the wasp.

During the war, Salsbury had sold his scooter business to a defense contractor firm by the name of Avion, a company that was soon absorbed by Northrup Aircraft Corporation. This relationship with the aeronautics industry led to the revolutionary design of the outrageous Salsbury Model 85, a scooter nearly as big as a car and as aerodynamic-looking as any Northrup jet plane. A perfect opposite of Cushman's Eagle, it was one of the first road vehicles ever to be tested in a wind tunnel. While the 85 can be credited for advancing scooter technology, it did not sell particularly well. Rather than credit Cushman for winning the first American scooter war, Foster Salsbury blamed the car for his downfall. "Demand fell off when cars started to become available again," he said, "and Northrup stopped production."

Harley-Davidson's Topper

Milwaukee's Harley-Davidson Motor Company is arguably the most significant and formidable motorcycle builder of all time. The company thrived throughout the decades despite very direct and vicious competition; first from Indian, its counterpart in the U. S., and later from British and Japanese competitors. It stood to reason that Harley-Davidson's motorcycle success might just translate to success in the scooter arena. In 1960 in Wisconsin, the land of the Cheeseheads, Harley-Davidson revealed its entry into the scooter marketplace, the Topper.

Looking at a Topper one might have been tempted to store one's cheese inside of it. The Topper looked like a refrigerator turned onto its side with a foam-and-vinyl seat, a boxy steel fender, and a bicycle-style handlebar. A rarity among scooters, its 165cc engine started with a rope, like a lawnmower, and it had an automatic transmission like a Cushman or a Salsbury. It was all-American, slathered with lots of shiny paint and targeted toward children. The popularity of scooters was falling like a stone by the time the Topper hit the streets, and kids might have seemed like the only consumers left buying them. Cushman, Salsbury, and even Lambretta had already begun winding down their stateside scooter operations. It follows that the Topper failed to live up to expectations. Today Toppers are largely a novelty among scooter collectors and enjoy greater value among Harley-Davidson enthusiasts, most of whom still consider them toys.

Vespa Does America, Allstate by Allstate

Aside from the lovely Ms. Hepburn, another American icon was instrumental in the introduction of the Vespa to the United States. Sears Roebuck and Company began selling 125cc three-speed Vespas in 1951 under their house brand, Allstate. America's prewar scooters had been loaded onto airplanes and shipped to Tuscany. Now Italian scooters were being loaded onto cargo ships in Tuscany and shipped to Sears stores across America.

Vespas were neither the first nor the only vehicles to be badged Allstate. Sears had a vision of selling all types of vehicles under the moniker. In 1951, two years after the death of Cushman president Charles Ammon, Sears began to sell Cushman scooters under the Allstate name. Several Allstate models were manufactured by Daimler Puch in the later 1950s and early 1960s including mopeds, scooters, and small motorcycles. Like the Cushman Allstates, these also sold very well. But North American collectors have a special place in their hearts for the Vespa Allstates, known as Cruisaire models, because they were the first of the new Italian scooters imported into North America.

The first Vespa Allstate, the Model 788.100, became available at Sears stores and through the company's famous mail-order catalog in 1951. The machines were timeless and adapted specifically for the American market. To keep costs down, Allstates were not equipped with some of the improvements that had been added to Italian Vespas by 1951. They lacked a front shock and were starkly decorated. Speedometers were not available; neither were passenger pads. The 788.100 was available in any color one wanted—so long as it was green. (Vespa is rumored to have used military surplus paint on the Allstates.) A luggage rack was mounted where Audrey Hepburn had sat, rendering the Allstate an "operator-only" model. The 788.100 featured one of the most beautiful headsets ever installed on a scooter: a cast aluminum headlamp clamped to the chrome handlebars, decorated with a spiny little ridge atop its teardrop profile. This was the first of many changes—some subtle, some not—that Vespa had to make in order to adapt to America's vehicle importation laws.

While Sears was also offering Cushman scooters under the Allstate brand, it was evident in the retailer's catalog that Vespa was the premium choice. They called the Cruisaire "our finest motor scooter" and tempted readers to "go continental with this fine Italian powerhouse." The Italian scooter revolution in America had begun. The retailing giant ordered 1,000 Cruisaires from Piaggio in early 1951; by September of the same year, 5,000 more were on their way. According to a 1952 report in *Time*, Sears planned to order 2,000 Cruisaires per month from that point on. At an advertised price of $279.95, the Cruisaire was a bona fide hit.

The model remained basically the same from 1951 to 1955, the year Vespa decided to open their own dealership network. Allstate models saw slight evolutionary changes between 1955 and 1966, including the addition of a traditional cast headset to replace the bicycle-style handlebars with which they first came equipped.

Sears also offered a four-speed model in 1964. It was badged an Allstate, but was faster and quicker than previous models. Sears remained involved with selling Vespa scooters until 1966, although for the last couple of model years the name Allstate was dropped in favor of, simply, Sears. These models from the mid 1960s, based on the Vespa Sprint 150 and the Primavera 125, are today referred to as "blue badge" models because of the metallic blue Sears badge that adorns their legshields and are highly sought by collectors.

Aboard Ship to the New World

The first officially imported Vespa models in North America were not Vespas, they were Allstates. Likewise, the first Lambrettas officially imported to North America were technically not Lambretta models, they were German copies built under license by NSU and imported by a New York-based BMW motorcycle distributor called Butler and Smith in 1951. Shortly afterward, two

distributors began handling the Innocenti-built models: Anderson Motorcycle Supply served the West Coast, New York's Baker and Shalit the East. Models available included the Model D and the brand new Model LD in both 125 and 150cc varieties. The D was as simple as a scooter could be, with very little bodywork, an abbreviated half-legshield, and a simple steel tube frame. While D models are very rare in North America (and highly coveted), they are not prime examples of Italian design like Vespas. Audrey Hepburn probably would not have been caught dead on one. Conversely, the LD model was much more appealing, with cleverly designed body panels hiding the mechanicals a lá Vespa. The LD's legshield was full-sized, also like a Vespa, and, unlike an Allstate Cruisaire, it was equipped to seat two.

By 1955, Innocenti established an American subsidiary. Now, instead of relying on distributors, the firm could select dealers and begin seriously competing with Sears, as well as the new dealership network Vespa was beginning to establish on their own.

Impressed with the success of the Allstate and itching to compete with Sears for a chunk of the American scooter sales pie, Montgomery Wards added scooters to its catalog and stores around 1957. Like Sears, Wards stocked its stores with scooters from a variety of makers. Lambrettas, Bianchi scooters from Italy, Sachs mopeds from Austria, and Mitsubishi Silver Pigeons were all sold under the house brand, Riverside. Like Allstates, Riversides were disguised as Wards' very own product. The attractive metal "Riverside" badges Wards used are considered a badge of honor on a vintage scooter. Next to the word "Riverside" appears a tiny letter "M" placed atop a tiny "W," together they stand for, you guessed it, Montgomery Wards. Like Vespa models were the cream of the Allstate crop, Lambrettas were the most desirable of the Riversides. By the late 1950s, Americans were convinced that Italian scooters were best.

In the mid to late 1950s, both Lambretta and Vespa were marketing scooters through mass merchandisers and through official dealerships. In 1955, Vespa pursued dealerships through a small network of American distributors and opened dealerships in almost every major city, as well as in small towns and rural areas. In the 1950s, motorcycles were still quite large and heavy and the ATV had not yet been invented. Farmers and other rural residents found that scooters worked nicely to visit the neighbors, check on the chickens, or simply ride to the mailbox. The fact that both Sears and Montgomery Wards offered scooters by mail made it easy for country folk to order one. The term "barn find" is often used to describe scooters from this era later pulled from farms, often with very low mileage and sometimes featuring fascinating makeshift repairs and alterations.

As the number of Vespa and Lambretta dealerships grew, the popularity of the scooters exploded across North America. Soon many of the big European scooter builders looked across the sea to expand their sales. Heinkels, Zundapps, Fujis, Mitsubishis, and many others were imported and sold to dealerships. Vespa and Lambretta dealerships often added scooters from other manufacturers in order to round out their offerings. Though they were encouraged to sell only proper factory models, dealers found they could keep scooter sales in the scooter dealerships and out of the motorcycle ones by adding a nice range of products. As a result, America and Canada established networks of scooter-specific dealerships, *scooter shops*, that waved the scooter flag high. Such shops encouraged consumers to show real enthusiasm for their scooters, rather than feel that buying a scooter was some sort of compromise. History would demonstrate that scooters were best sold in a setting of their own, apart from the bigger bikes. The macho, gritty environment of a motorcycle dealership tended to alienate scooter buyers, and the scooters were somehow always dusty.

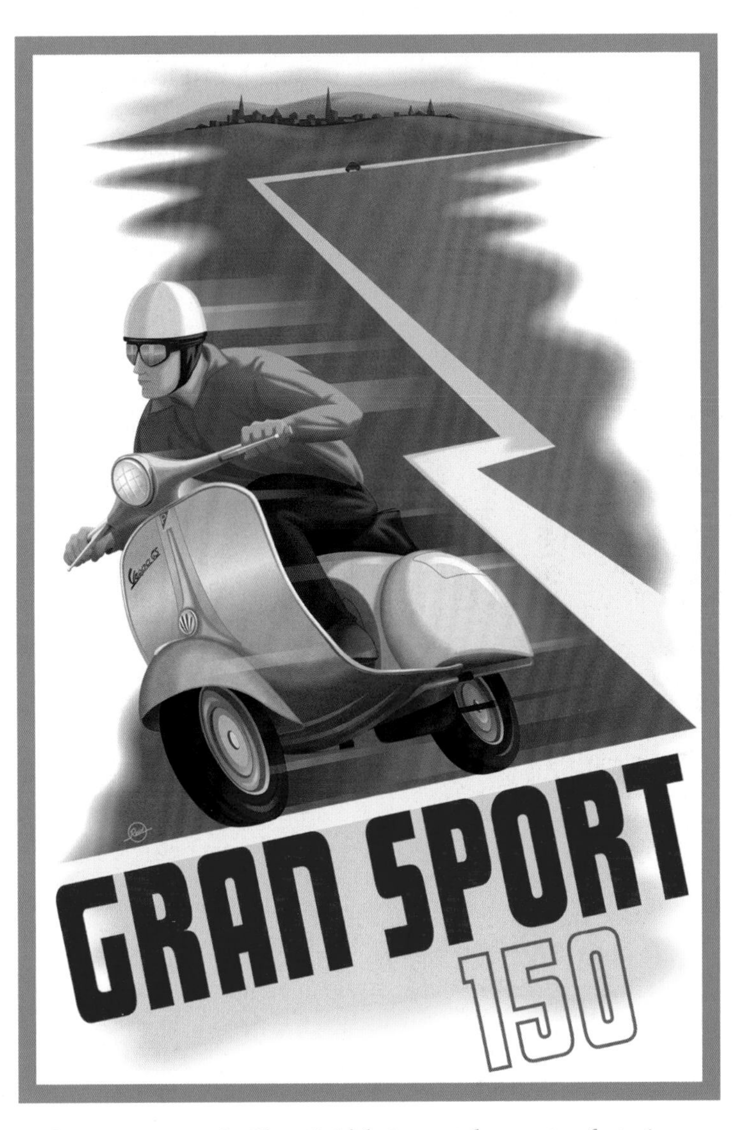

Contemporary artist Glenn Reid designs art deco posters featuring classic scooters

Young parents, Mr. and Mrs. Hal Emmons rest on their scooter with their baby, Kathy, on the floorboard. The family rode the scooter from Miami to Chicago in five days, 1961

If You Can't Beat 'Em, Join 'Em: Cushman Goes Continental

As wave after wave of shiny new European scooters crashed onto the shores of the North American continent, Cushman Motors in Lincoln just kept plugging along. Throughout the mid and late 1950s, the Eagle was the cool scooter to have in many circles. The big-bike-like little bikes sold in impressive numbers. All over the Midwest, old timers will tell you all about how they terrorized the neighborhood on an Eagle when they were fourteen years old. Eagle models remained in production up until 1965 when Cushman finally threw in the scooter towel and devoted itself to other product categories like golf carts.

Cushman and Cushman/Allstate step-through models, such as the RoadKing, did not sell terrifically. While these machines from the late 1950s were gorgeous, they had a hard time competing with the steadily improving Italian machines.

In 1961, Cushman finally sold its soul. Deserting all but the Eagle series, relegating its step-through models to history's dustbin, Cushman began distributing Vespas through its very own dealerships. American consumers now had three ways to shop for Vespas: Sears, authorized Vespa dealers, and the Cushman dealer network. (It is not difficult to understand how this particular circumstance was not terribly popular with Vespa dealers!) Cushman then licensed the venerable Gran Sport 150, one of the best Vespa models ever built, and the first to wear ten-inch wheels rather than eight-inch ones, and offer four speeds rather than just three.

Available in Europe in 1955, the GS150 had opened the floodgates of scooter development a few years earlier. The first true performance scooter, its advertised top speed was sixty miles per hour. The GS150 was as fast as all but the largest motorcycles of the day, but only 500 were distributed by Cushman in the early 1960s. Today, GS150 models with the "Cushman Motors, Lincoln, Nebraska" logo on the legshield can fetch up to $10,000 in restored condition.

Scooters Get Fast, Sales Slow Down

With the GS150, Piaggio had launched a warning flare

toward Innocenti. The era of the hot rod scooter had begun. Scooters were no longer utilitarian putt-putts. Innocenti blasted back a salvo of its own with another of history's most incredible, almost edible, scooters, the Tourismo Veloce 175. The TV175 was more than a response to Piaggio's challenge. After the hiccups of Series I, Series II and III TVs were game-winning plays, one-upping the Pontedera plant.

Soon the GS150 would be followed by the GS160, then the SS180 and, finally, the Rally 200. Innocenti's engine sizes increased right along with Vespas'. The TV series was developed and perfected between the mid 1950s and the mid 1960s. The venerable Tourismo Veloce would later become the first production two-wheeler to feature a front disc brake. This was a drastic improvement in stopping power and is still considered a high-end feature today. Vespa did not offer a disc front brake on any scooter until the late 1990s.

Innocenti topped its TV models in 1966 when they released the X200 Special, better known as the SX200. Considered by most scooter collectors to be the very finest Lambretta ever built, the SX200 raised the bar to the very top. Innocenti would never eclipse it.

American consumers often had to wait a year or two before they could get their hands on the latest scooters out of Italy. Dealers had to clear out older inventories and distributors waited to win approval from the government to import current models. It is for this reason that the titles for vintage scooters often reflect the incorrect model year. A 1958 GS150 might have paperwork that reads 1963.

Italian scooters finally forced American ones almost completely out of their own market simply by being a better value, not necessarily by being cheaper. The initial sales surge spawned by their sudden availability had slowed and sales went from super duper to simply steady. Sears and Wards bowed out of the scooter arena not long after Cushman did. One must assume that it simply was not profitable for them anymore. Still, the scooter market in North America could support a small network of dedicated European scooter dealers.

The 1960s were a famously tempestuous time in North America. America's moments in the sun were over and Americans were through rewarding themselves.

Many North Americans travelling in Europe found it convenient, inexpensive, and fun to use scooters as their transportation. Here cousins Sheila and Elaine, and their trusty Vespa, nicknamed Francoise, spent three and a half months enjoying Europe in 1958.

Once again, scooters had to justify their cost by performing their duties in an efficient and reliable manner. Social change and civil rights and war and peace were on the minds of most folks. Tailfins and chrome trim were giving way to bare-bones, austere, stripped-down family sedans with gargantuan, fire-breathing 350 horsepower mega engines. "Muscle cars" they called them. Tiny European and Japanese cars, the likes of which America had never really had any use for, began to catch on, too. For Piaggio and Innocenti, and the dealers who sold their scooters, the sixties were relatively uneventful in terms of sales despite a steady stream of exciting new models. Scooter shop owners earned a living but nobody was getting rich selling scooters in America during that decade.

Like Oil and Scooter: A Shot in the Arm Courtesy of OPEC

In the early 1970s, America had to come to terms with something a bit disturbing: It had become addicted to that petrochemical fuel farmed from the dry deserts of the holy land and refined to a smelly clear liquid called gasoline. With precious little control over oil supplies, the land of the free and the home of the brave was brought to its knees when, in 1973, the oil producing nations of the Arab world placed an embargo on oil exports to the Western World. Furious over the support of the West for Israel, Arab nations decided to hold out on some of their best customers—Americans—in order to intensify pressure on the world to pull out of the Middle East. The Organization of Oil Producing Countries, also known as OPEC, oversaw the embargo. America suffered from severe withdrawal symptoms when a massive fuel shortage ensued, forcing long gas lines and a sudden scramble to conserve fuel. Instantly Americans were forced to reconsider their attitudes toward energy. Gone were 455-cubic-inch engines and hood scoops. A new buzzword was circulating around America: *conservation*.

One very obvious selling point on scooters was, and always had been, excellent fuel economy. Even in the opulence of the 1950s, scooters were advertised to be efficient and inexpensive to operate. Vespa had offered 100 miles to the gallon decades before Americans came to the realization that gasoline would not always be cheap and easy to get. It did not take many two-hour waits in gas lines before scooters began to make very good sense to the consumer. Automobiles had reached such excess that their average fuel mileage was below ten miles per gallon. Scooters offered ten times that. The little machines began to draw attention from the media and, again, appeared regularly in movies and advertisements. They were quickly returning the national consciousness after a decade or so of relative obscurity.

The American Vespa dealer network began to flourish like it never had before. By 1973 Sears and Wards had long since stopped selling scooters. Innocenti had fallen on hard times and Italian Lambretta production was stopped. This left the American market absolutely ripe for the Vespa. With the exception of skinny, smoky little mopeds from France, Austria, and Italy, which were cheap but nerdy and hard to use, the new surge of scooter shoppers were left with very few alternatives besides a visit to their local Vespa dealer.

In 1973, Piaggio was offering a very solid range of Vespa models. Spared from a great deal of competition, Piaggio slowed the pace of new model releases for a period in the later 1960s and early 1970s. In fact, most of the models offered in North America were cast-offs from the European market, each fitted with a new, block-letter style legshield badge rather than the script-style Vespa logos of the past. Two examples of this are the Primivera 125 and the 150 Super, each had been offered in Europe only into the mid 1960s, yet were sold in the USA for almost another decade.

To commemorate the 50th anniversary of Dave Lee and Rick Ronvik's (above) 1953 trip by scooter through Europe, Genuine Motor Company sent them back to Europe on brand new Stellas

The model line up evolved very slowly and remained mostly the same from one model year to the next during the mid 1970s. North American consumers could chose from the 50cc, three-speed 50 Special; the three-speed Vespa 90; the 125cc, four-speed Primivera; the 150cc Sprint; the high performance Sprint Veloce; the 150 Super (based on the Sprint but with eight-inch wheels instead of ten inch); and the awesome Rally 200, the fastest Vespa ever built and the first to be offered with electronic ignition.

Sales reached record levels and the Vespa dealer network grew. Suddenly, scooters began to appear in big cities and in little towns, much like they had in the early 1950s when Sears first began offering the Allstate. Today, the blocky-badged sixties-tech scooters sold in the mid 1970s are among the most prominently seen scooters at rallies and events. Scooter kids have been hauling them out of Grandma's back yard and Farmer Brown's barn since the early 1980s. Thanks to the sheer numbers of them sold, the 1970s Vespas were everywhere, and still are. Piaggio was not able to rest on their laurels in the North American market for long, though. The Japanese were coming. Things (particularly scooters) were about to get ugly.

One Last Encore: The P Series and ... Curtains

It took America a little while to recover from the tough times brought by the 1970s. Americans were still stung just enough to shell out for a "K car" or some tiny Mazda with awful build quality because they figured it didn't use too much gas. Since they were easily the most miserable years in the history of the American auto industry, the early 1980s were the ideal time for Piaggio to introduce an innovative new scooter with all of the improvements it needed to be considered a truly viable alternative to a lame car. For once North America would not wait for the newest Vespa; the P Series was shipped immediately.

By 1978 red ones, white ones, orange ones, dark blue ones, light blue ones, tan ones, maroon ones, and silver ones would buzz out of dealerships in numbers North American Vespa distributors had not seen before. Also known as the "P-2-Common" due to its incredibly strong North American sales, Vespa's P200E might be the greatest Vespa of all time, or the worst; it's a matter of perspective. Depending on who you ask, the larger of the two models in the P Series or *nuova linea* (new line), the P200, was Piaggio's workhorse or the ugliest pile of garbage eyes have ever lied upon. No one will deny that the P was fast. The 200cc engine, only slightly modified from that of the Rally 200, was extremely strong and produced 12 to 13 horsepower. No one will deny that it was reliable. Its modern twelve-volt electrical system with electronic ignition was the finest Piaggio had ever designed. Also no one will deny that it was comfortable. The ergonomics were quite nice, with a thickly padded seat, convenient controls and switches, and plenty of room for a passenger. Furthermore, no one will say that the P could not be personalized. No other scooter in history had more accessories built for it. P owners could get everything from massive luggage boxes to chrome crash guards to sidecars and FM radios. Where the controversy lies is the P Series' boxy 1980s styling. It was a radical departure from traditional Vespa designs. Enthusiasts were very put off by the new top-of-the-line Vespa. Piaggio was an innovative company, though, and they were not afraid to push the boundaries about once every ten years or so.

The very first 200s were badged "1977" and share some features with the Rally 200 that subsequent models did not. Also available was the P125X. Both sold extremely well. They were marketed to the young and purchased largely by the middle-aged. Piaggio advertised the P Series heavily in major publications like *Playboy*. These venerable scooters are by far the most commonly found today. Many a scooter boy and girl has started their hobby by buying a P Series Vespa.

The P Series Vespas were popular, reliable, and just right for the oil-deprived North American market

The P200E and P125X were sold in the USA from 1978 to 1983 in an official capacity and with Piaggio's involvement. Scattered imports of P Series Vespas, including the electric start PX150 continued well into the late 1980s, but they were not officially sanctioned. They were imported through the grey market. Piaggio was forced to pack up and completely leave the USA thanks to a new round of stricter emissions and safety standards, which basically outlawed the importation of big two-cycle scooters. Since the company did not have any products to offer that complied with the new regulations, it simply folded its American division. Dealers were left with scooters on the floor, parts inventories, curious customers, and a big mess to clean up. No longer could they pick up the phone and get assistance from Piaggio. It was gone. One by one American Vespa dealers liquidated and shut their doors. The stuff they left behind would become the initial inventories of the next generation of American scooter shops.

Japanese Occupy the American Scooter Market

With the untimely exit of Piaggio from the American market came the rebirth of the Japanese scooter industry in the U. S. Honda and Yamaha had already taken a serious bite of the Vespa and Lambretta market with their super-high-quality, small-bore motorcycles and their smoky mopeds. Now the scooter market was theirs, all theirs.

In 1983 Honda released their Aero models, the Aero 50 and the Aero 80. Both were two-cycle models with decent power, automatic transmissions, and styling that made scooter enthusiasts gag. The days of manufacturers catering to those types of customers were over, though. Honda had little interest in clubs or rallies or any of that. Scooters were built in Japan to be cheap and efficient, characteristics that scooters had always possessed but never in such a pure form, so lacking in personality. Along with the two-cycle Aeros, Honda offered their Spree 50, a stripped down, even cheaper two stroke, as well as the Elite 50, 125, 150, and 250. Each of the larger models featured a four-cycle engine per the regulations of the day. No two-cycle models besides the Aero 80 were offered and it was only available for one model year. The four-cycle Hondas were seamless and as reliable as the day is long. Honda scooters almost never broke down.

In the garage at Chicago's Scooterworks

Yamaha plotted a similar course to Hondas. Their offerings included the Riva and the Razz in several different configurations and engine sizes. Thanks to the Japanese manufacturers' excellent dealer networks, Rivas and Razzes and Elites and Sprees were sold much more widely and effectively than Vespas or Lambrettas or Cushmans ever had been, and the sales were enormous, beyond anything the other scooter manufacturers had ever dreamed of. In the 1980s, urban motorcycle dealerships were selling more scooters than they were selling motorcycles. Even though these models were painfully ugly and offered no real value to collectors, they remained available and relatively unchanged for more than fifteen years. From 1983 to 1997 the Japanese owned the American scooter market outright. There were simply no other credible products available.

Stayin' Alive: Keeping the Old Ones On the Road

After Vespa's departure from the North American market, a massive void remained for enthusiastic scooter owners seeking standard parts and service. Most of the original Vespa dealerships closed up shop and sought opportunities in other arenas. For example, Denver Vespa became Scooter Liquors. The inventories of parts and scooters that remained were simply a liability. They were mostly auctioned off, wholesaled, and gotten rid of in the most efficient possible way. A handful of former dealers held on tightly to their stashes and tried desperately to stay in business, but the majority of them simply walked away allowing the second generation of American scooter shops to develop.

Entrepreneurs fill voids. Enthusiastic young scooter fanatics were quick to start snatching up those burdensome excess inventories of scooters and parts. They believed that people who loved Vespas and Lambrettas would continue to love them regardless of the brands departure from the marketplace. They were right. And so, in the middle of the "Scooter Ice Age" a handful of small, independent businesses assumed the responsibility of servicing and reselling old scooters. Most of these scooter fix-it shops were in California where the largest numbers of scooters were originally sold. Some specialized in repair, restoration, and customs. Some keyed in primarily on the parts end of the business, assembling mail-order catalogs and relying on the UPS man to deliver the goods on time. At-home repairs became more common in the absence of dealerships. Soon the mail-order shops were unable to rely on surplus inventory alone and they began importing parts directly from Italy. Shops like Scooterville in Anaheim, San Diego's Vespa Supershop, San Francisco's First Kick Scooters, and New Jersey's Scooters Originali were the keepers of the Vespa flame.

In Chicago an aggressive and book smart opportunist hit the mother lode. He stumbled upon scads of brand-new, still-in-the-crate Vespa P200Es stored in a warehouse space. The man who owned them was eager to sell and he was ready to buy. He then started buying Vespas at auctions, from classified ads, estate sales, and former dealers. He sold them as quickly as he could find them. He also assumed the inventory of a Chicago based Vespa distributor that included all types of parts and accessories. When it all became too much to do from his garage, he rented a small, funky building on a residential strip of Ravenswood Drive in Chicago. Scooterworks USA was born. Its founder, opportunist Philip McCaleb, pursued the goal of, as he likes to say, "Creating something from nothing."

The Scooterworks crew.

A religious Bears fan, a devoted father, a lover of hot dogs, and a sucker for a smart deal, McCaleb became to scooters what Don King was to boxing; a consummate and tireless promoter. He placed advertisements in publications large and small, both selling and seeking scooters and parts. He also paid personal visits to important European parts and accessory suppliers in order to arrange exclusive importation agreements. Nobody devoted the energy or the expertise to the business of scooters that he did. McCaleb drew the admiration and scorn of scooterists both stateside and overseas. His fledgling mail-order catalog, filled with items for Vespa scooters from every era, became the defacto pricing guide for many items, which cut into the profits of smaller shops who were unable to buy in the massive quantities Scooterworks could.

As Scooterworks' mailing list grew, and McCaleb became a master of data management, scooters left for dead around America were risen. His catalog reached tens of thousands, then hundreds of thousands of households. Offering both parts and manuals to show people how to install parts, the catalogs became like bibles to scooterphiles. Unlike most scooter shops where the main tools were wrenches and screwdrivers, Scooterworks' main tools were telephones and package tape dispensers. With the help of Scooterworks, as well as the more well-established California shops, Vespa enthusiasts were able to keep their old scooters on the road. As a result, scooter clubs gained momentum and the rally scene began to grow by leaps and bounds. Still, the scooter market was limited by the lack of new products.

In the late 1990s, importers rushed to gather scooters from Europe, stuff them into sea crates, and resell them in North America. Italy offered incentives to scooter owners to upgrade to cleaner, less polluting scooters and offer their old ones for destruction. Instead of being destroyed, the tired machines were auctioned off to the highest bidder. Often that bidder intended to export them either to North America or to Japan, where vintage scooter sales were also booming. Outfits like Pirate Imports in Austin, Texas, and Scooterworks in Chicago, shipped hundreds of scooters back to the States. They were sold to private parties and also wholesaled to other scooter shops. Since these were generally very high mileage scooters in need of extensive repairs, parts sales surged. Eventually the well began to run dry. Decent vintage scooters became harder and harder to find; even in Italy.

The Next Italian Renaissance: Enter the Velocifero

The market was ripe for new European scooters by the late 1990s. With Honda and Yamaha perched comfortably atop the sales charts and most European models barred from the North American market place for well over a decade, conditions were perfect for a sort of scooter consumer uprising. While interest in the boxy, plastic Japanese "same old, same old" had long since waned, and motorcycle shops ordered scooters only out of a sense of obligation, vintage scooter sales were running high. Values of older models were skyrocketing and hard working little scooter shops were thriving; indicating not just that consumers were itching to ride scooters again, but that they were interested in *being seen* on scooters again.

In 1997 a gutsy New York couple, Joel and Susan Sacher received approval to import and sell a scooter worth being seen on: the artsy Italjet Velocifero. Already insanely popular in Europe, the 50cc Velocifero was retro chic and ready for market, it only needed a proper escort. Several individuals had "smuggled" grey market Velociferos into North America, but nobody had committed to import them formally. Because of its small displacement and efficient two-cycle engine, only moderate changes needed to be made to the design for the scooter to meet American standards. Its popularity was largely due to its familiar Vespa-style steel legshield and floorboard section, which was supplied by India's Bajaj, a former partner of Piaggio and Vespa's parent company. This piece was grafted onto a tubular steel subframe that was concealed, along with the majority of the mechanicals, beneath a beautifully molded plastic tail section.

The Sachers took a huge gamble by providing the capital to homologate the Velocifero and by purchasing the massive product liability insurance policy required in order to attract legitimate dealers. They named their new company Italjet USA, after the Italian manufacturer for whom they were distributing. They ran the company from their Manhattan BMW motorcycle dealership, BMWNY. The New York City motorcycle dealership instantly became a national distributor with far reaching responsibilities: warranty claims to process, credit terms to extend, parts inventory to keep. The immediate and thunderous success of the Velocifero turned out to mean a whole lot of work for the Sacher family.

Dealership applications burst in to Italjet USA from all over the country. The company was bludgeoned with requests from small scooter shops, huge motorcycle dealerships, even small engine dealers, recreation vehicle outfits, and luxury car lots. It seemed as though a whole lot people were excited about the first Italian scooter introduced in America since the Vespa P Series in 1977.

Many of the dealerships the Sachers selected were scooter shops that had survived almost exclusively on repair work up until the introduction of the Velocifero. Living in constant anticipation of Vespa's return to the states, these keepers of the flame were the main reason that old scooters still held appeal. The problem was that they had never sold new products before. Enter the Sachers who helped turn scooter shops into scooter dealerships. Soon Italjet USA introduced new models and the Itlajet brand was complete. The Formula, the Torpedo, and the Dragster, along with the venerable Velocifero, offered scooter consumers their first taste of the scooter good life in a long, long time. A bold and brash New York couple created a brand new scooter industry. Scooter manufacturers from Europe and Asia watched their experiment with great interest.

A New Millennium, Another Swarm

Even with the relative success of the Velocifero, the American scooter market was hardly a blip on the radar by 1999. One major Italian manufacturer, however, decided to open a wholly owned subsidiary in the United States and begin widely distributing some of the best scooters in the world through a large dealership network. Regarded by sport motorcycle and racing enthusiasts as an almost mythical company, Aprilia was heralded for its fantastic big bikes and massive trophy case filled with the spoils of multiple road racing

American Scooterist, *the newsletter of the Vespa Club of America*

The elegant Velocifero gave fans of retro styling a modern scooter to choose

world championships. Despite its high-speed heritage, Aprilia's bread and butter products were scooters. One model, in particular, the Scarabeo 50, was one of the best selling in all the world.

Since most of the hype surrounding the company related to racing, motorcycle dealerships in the USA who took the line on were not responding well to the scooters. Aprilia required that the dealers buy a small number of them and that is exactly what most of them did, accepting only the minimum order. Aprilia's sales staff noticed, though, that scooter shops with established customer bases were calling regularly begging to visit with them. These small businesses did not meet the minimum financial requirements in most cases to set up the credit lines needed to stock a dealership full of shiny motorcycles, and the scooter line was not available separately. Most could not even swallow Aprilia's minimum orders for parts, accessories, and signage. To remedy this problem, Aprilia cleverly crafted the designation of "scooter-only dealerships," allowing business people with less capital to jump on board. The success of Aprilia scooters was instant, leaving the company with far fewer scooters than the market demanded. Scooter-only dealerships led the charge, outpacing even large motorcycle dealerships in unit sales. Aprilia had discovered what many already knew: scooter people sell scooters.

In addition to the Scarabeo 50, Aprilia introduced the race inspired, water-cooled SR50, as well as the mammoth Scarabeo 150, within just months of each other. In the meanwhile, government testing facilities were busily handling approvals for dozens of new models from Europe. Derbi, Malaguti, and Kymco, a Taiwanese firm with very strong European sales, all introduced full lines and scrambled to secure the best dealers. With the introduction of these brands, American consumers could choose from five different manufacturers of Europe's most popular scooters. It was a sudden and wonderful end to the "Scooter Ice Age."

Return of the Wasp

Rumors about the return of Vespa had swirled for more than five years by the time it actually happened. If not for complications relating to the trademark, Piaggio might have strategically beaten to market the brands that were already flourishing early in the new millennium. As it turned out their timing was perfect. The anticipation was palatable by that time and the availability of scooters had only whetted the consumer's appetite. Piaggio, like Aprilia, established a wholly owned USA subsidiary rather that relying on a network of distributors the way they had the first time around. They also employed a controversial strategy on selecting dealerships. The company hired a New York based marketing firm to pound the pavement looking for operators to open Vespa boutiques, very high-end retail stores selling nothing but Vespa and Vespa "lifestyle" items, such as $50 tee shirts and $400 rubber jackets. Apparel and other items such as colored pencil sets, fancy soaps, and pricey watches were a critical part of the equation for boutiques. The idea was to work so hard branding the name Vespa that it became bigger than scooters; a fashion marque with power to sell products of all types.

An expatriate from the Italian offices of Piaggio joined forces with a notoriously egotistical marketing guru named Peter Laitmon and scoured the country for wealthy investors willing to stake nearly a half a million dollars per store. Each was to be nearly identical. Several formats were offered in catalog form for the potential operator to consider, including a kiosk intended for placement in airports. All interior fixtures were to be purchased from Piaggio along with all promotional materials and, most importantly, all parts and accessories. The clothes, the soaps, and the pencils were all required inventory and the quantities to be purchased were at the discretion of Piaggio. This approach to dealer selection was unique to say the least. Most manufacturers approached existing shops as potential customers. Piaggio sought to open its own chain of franchise-like exclusive dealerships, none of whom offered any other scooter products. Unfortunately for Piaggio, the experiment did not work. Only a handful of boutiques opened and the company was forced to look for other ways to sell scooters while not leaving those who did invest in the boutique concept

The exquisitly styled Vespa ET2 and ET4 made the return of Vespa to North America a success and have helped push scooters sales of all brands to historical highs.

furious over the declining commitment necessary to become a dealer. Several jumped ship after hemorrhaging cash for the first two years.

Piaggio's somewhat snooty return to the market left a sour taste in the mouths of most of the nation's highest volume, most experienced scooter shops. In fact, Piaggio had petitioned them to sign a contract "authorizing" them to service vintage scooters, an act they had already been performing for over a decade. In return the shops were to receive parts support and recognition on the Piaggio web site. The parts never came and the web site was soon redesigned without the listings. The contract, called the "Vintage Vespa Restoration Shop" (or VVRS) agreement, was really nothing more than an aid to the company in their trademark disputes. The signer was essentially agreeing that Piaggio retained all right to their trademarks and that it had never abandoned the U. S. market. It offered virtually no benefit to the shops whatsoever.

Shady dealings like this typified and needlessly soiled Vespas much heralded return. Collectors and enthusiasts greeted Vespa with open arms only to be told that they were not the new target market. Piaggio held black-tie galas in New York and L. A. to celebrate their products. Invited were A-list celebrities and wealthy socialites. Ignored were the filthy masses of scooter riders who made up the largest segment of potential buyers. In the year of Vespa's return, the Vespa Club of America's national rally, known as Amerivespa, was held in San Diego. No representative of Piaggio was present even though it was held in the home state of the corporate office.

Vespas sold extremely well despite the feelings of the hardened enthusiasts. Two models were introduced: the 50cc ET2 and the 150cc ET4. Identical in appearance, the scooters were advertised to have top speeds of thirty-five miles per hour and sixty-five miles per hour. Well constructed and skinned in steel, they exuded quality and were styled perfectly. The color selection was far and away the best in the business and Vespa showrooms shone like gleaming bags of jelly beans. Vespa dealers did not struggle at all with the fact that Vespas were much more expensive than other scooters; they touted it as a sign that they were better. Buyers could choose from many beautiful accessories to make their Vespas stand out from the rest. Piaggio marketed their Vespas very heavily and mastered the art of product placement, prompting Vespas to appear in movies, on television, and in commercials. As the populous became more aware of Vespa they became more aware of scooters in general and sales of all brands began to soar.

Cheaper Chic: Honda, Yamaha, and the Sincerest Form of Flattery

It only took a couple seasons of sales growth in the new scooter market before the Japanese heavyweights decided to mix up their product offerings. Yamaha introduced a cute and stylish little two-cycle 50 called the Vino in 2001. The name was appropriate since the Vino was an obvious clone of an Italian machine. Priced well below $2,000, it immediately began to strip sales away from the European manufacturers despite the fact that it was too small to carry a passenger. Yamaha demonstrated that it was capable of producing appealing scooters at a price point only another Japanese manufacturer could match. The following year Honda retaliated with its fun, but painfully underpowered Metropolitan with the first four-cycle engine among the modern 50cc scooters. Like the Vino, the Metro featured styling cues clearly lifted from scooters like the Velocifero. It was cute and familiar looking, with trademark Honda fit and finish.

During the decade of the 1990s, big Japanese motorcycle companies worked overtime trying to reproduce the look and feel, even the sound, of Harley-Davidson motorcycles in order to capitalize on the world-wide popularity of the famous American "big twins." In the new millennium they sought to do the same with Italian scooters. Because of their enormous dealership networks, Yamaha and Honda scooters were far and away the best sellers in America almost as quickly as they were introduced. Consumers in smaller markets and those not living near traditional scooter shops finally had access to scooters that looked just like the ones they kept seeing on television, even if they were just knock offs. The quality of the machines was not a concern as both manufacturers had sterling reputations. Unfortunately, the machines were somewhat bland and unexciting entrants into the 50cc genre.

The large displacement "maxi scooters" that began to arrive from Japan at around the same time offered a much larger dose of excitement. The Honda Silverwing 600, Honda Reflex 250, and Suzuki Burgman 650 and 400 were truly innovative scooters based on the European notion that scooters were more than just urban transportation. Maxis were designed for travel between towns on the high-speed freeways of Europe. Scooters that easily exceeded 100 miles per hour were unheard of in North America prior to these models being released. In this instance the Japanese beat the Europeans to the table.

Aprilia released its amazing Atlantic 500 several long sales months after the Japanese models had already hit showrooms. Gigantic maxi scooters were somewhat limited in appeal since Americans were unfamiliar with the product category. Maxis sold at a price point where consumers could easily choose a motorcycle. Sales started out slowly. As more and more people began to try the big models, though, sales picked up. The smooth power and effortless ride of these "Cadillacs" had enormous appeal to people for whom a touring motorcycle, like a Honda Goldwing, was simply too large. Maxi scooters offered all of the weather protection, comfort, and speed of a big Wing but at a fraction of the weight and a fraction of the price. Automatic transmissions meant carefree touring and a focus less on riding and more on the ride. Recreational riders and commuters alike found the versatility of maxi scooters refreshing.

India Offers Traditional Scooters

Piaggio partnered with two Indian firms to build Vespas through the years. Those companies were called Bajaj and LML. When their partnerships with Piaggio were through, both were able to continue building Piaggio designed products thanks to favorable rulings in Indian courts. In the new millennium Bajaj began offering four-cycle 150cc scooters with manual transmissions in the USA through a San Francisco based subsidiary called Bajaj USA. Consumers loved the fact that these scooters shifted gears and were built from solid steel. The Legend 150 and the Chetak 150, Bajaj's two USA models, were the first cush scooters offered in the States in nearly twenty years.

Genuine Motor Company's Stella has all of the allure and glamour that classic scooter enthusiasts crave

Philip McCaleb, the founder of Scooterworks USA in Chicago, approached the LML factory about importing their scooters as well. McCaleb faced a larger challenge, however, than Bajaj had. The LML traditional models were still powered by two-cycle engines, which did not meet EPA emissions standards. He had to invest loads of cash in the development of a catalytic converter for the scooters. Thankfully, his efforts to make the scooters comply were successful. He then selected his own paint colors, added several upgraded components to LML's existing scooter, and created the wonderful Stella, a precise reproduction of the Vespa P Series scooters with a snappy two-stroke motor and a familiar four-speed transmission.

Chinese Scooters: Onto Every Parade a Little Rain Must Fall

Just when the scooter industry began to ride high again for the first time in decades, unscrupulous importers began distributing on a massive scale cheap, poor-quality scooters made in China. Honda and Yamaha had dropped the ticket price to the scooter show by almost half, offering scooters for well under $2,000. These shady characters sought to undercut the big Japanese companies by the same margin, offering scooters that looked exactly like those from the Land of the Rising Sun for well under $1,000. In fact, the Yamaha Vino was copied time and time again by factory after factory. Yamaha, having licensed the design of the Vino, could do nothing to stop the importation of thousands of scooters identical to the ones their dealers were selling. The quality of these machines varied widely but the majority of them were awful. Despite their resemblance to Japanese scooters, they did not stack up. Cheap, spindly parts were used in their manufacture and, as a result, reliability was often extremely poor.

No fewer than twenty different brands emerged by 2004, including names like Geeley, Zhejiang, Qingqi, and Taizhou Kaitong. Rather than suffering such formalities as formal DOT approvals, product liability insurance, and dealer network selection, these manufacturers found every way possible to move their scooters in by the thousands while flying beneath the radar. Chinese scooters were sold on price alone, usually by the container load and usually to a reseller with no experience whatsoever in the business of selling scooters. The new breed of "scooter shops" operated over the Internet. They listed scooters on online auction sites by the thousands and wholesaled them to schlock used car lots and other unqualified resellers. Scooters showed up at flea markets and on street corners like bad Elvis rugs. Since most of them lacked DOT approval for sale as motor vehicles (and the needed lighting changes and emission controls) they were sold for "off-road use only," despite the clear intention of the buyers to operate them on the streets.

Mass merchandisers even jumped on the Chinese scooter bandwagon. Oddly, auto parts stores took a particular interest, offering both gas and electric models for sale in their stores. Costco, a discount members-only mega-retailer offered the Twist-N-Go brand Venice 50 to its members for $699. The Venice, a good quality, DOT legal Vino copy, sold very briskly. Unfortunately, Costco's liberal return policies and lack of repair facilities meant each store filled up with broken and leaking Venices. Traditional scooter shops were brought in to serve as "authorized repair shops." They hauled the broken scooters out of the Costco stores, fixed them, and sold them again. When Costco finally pulled the plug on what had been a costly foray into the scooter business, the repair centers became the dealerships. Twist-N-Go's brand image immediately improved.

Cheap Threats

Low grade, low priced products are the greatest threat to the scooter industry in North America. Should the public's price perception fall to the levels offered by these products, legitimate retailers will find themselves in a real fix. It is impossible to stomach the overhead of a full-time shop while selling scooters for $600 a piece.

Government controls intended to weed out cheap, non-DOT approved scooters must be rigorously enforced. European scooters will otherwise be priced out of the market in much the same way European motorcycle makers were when Honda introduced their bikes in America in the early 1960s.

There is also the concern that public confidence in scooters could be badly shaken if too many shoddy units are sold, leaving consumers to choose other alternatives. It is important that scooters that are sold are ridden regularly, not parked in garages and left for dead. Full-service shops must offer scooter consumers convenience and quality service if they are to be sustained. Otherwise, motorcycle dealerships are likely to swallow them whole.

Another looming threat to the scooter market lies in the event of changes in emissions regulations in America and Canada. A nationwide ban on two-cycle engines is in the pipeline, and unless emerging technologies can make a significant and rapid impression on decision makers, many of the models sold today will disappear. Manufacturers are working closely with regulatory agencies in an attempt to rescue the two-stroke engine from history's scrap yard.

Movin' On Down the Road: The Future of Scooters in North America

Scooters are undoubtedly here to stay, even if they go through future scooter ice ages. Yet nagging questions remain as to the sustainability of the current popularity they have enjoyed since 1998. Sales figures have exploded over the years, but it is doubtful that the rate of growth will continue. Most experts agree, though, that, regardless of the long-term outlook, the market will experience strong growth for anywhere from five to ten more years. Not only does the selection of scooters continue to expand but other factors—demographics, urban renewal, gas prices—contributing to their popularity remain favorable.

But aside from the economical future of scoots, the real question is what can we look forward to in the way of styles and production? Several trends will emerge as scooter builders continue to make bold innovations both mechanical and cosmetic. As options increase, consumers will have a more difficult, yet much more exciting, time choosing the scooters that are right for them.

Styling

Scooters will become more and more angular and racy as retro goes back into the past and performance becomes the order of the day. Look for sport bike-like innovations such as exposed aluminum frames, radical suspension systems, and even more powerful brakes in packages that closely resemble their larger motorcycle cousins. Designs that once seemed ultra-futuristic will become commonplace as manufacturers participate in the grand game of one-upmanship.

A New Lambretta?

Importers of the Twist-N-Go brand of scooters will reintroduce the Lambretta brand to compete with the Vespas and the Stella for the hearts and minds of vintage scooter enthusiasts looking for modern thrills. The new Lambretta will be designed under the supervision of Lambretta lovers from the U. K. and will be assembled in Italy. Styling should be reminiscent of the classic 1960s LI models.

The reintroduction of Lambretta is certain to impress fans of the marque and push the style factor of all scooters to greater heights

More and More Maxis

The Maxi scooter revolution will continue with more and more models being offered with powerful engines larger that 250ccs. Yamaha will enter the fray, introducing their Majesty models, which have been very popular in Europe. The displacements of these machines will continue to climb well beyond 600ccs. Motorcycle enthusiasts will become far more likely to plunk down their hard-earned cash on scooters.

Better Quality Scooters from China

Due to very low labor costs, scooter manufacturing in China will increase and exports will rise. Major manufacturers from all over the world will begin to move their manufacturing there. Chinese brands will be imported by quality distributors who will more carefully appoint dealers. Parts and warranty backing will improve along with fit and finish and overall quality.

Direct Injection

Direct injection technology is a sort of fuel injection that allows an engine to combust and re-combust the same fuel charge over and over again, spending more of the polluting gasses that would otherwise be expelled. Orbital Technologies, a company specializing in direct injection, has sold their systems to scooter manufacturers around the world. Aprilia, the first to install the system on scooters bound for North America, has enjoyed great success with direct injection. Their SR50 and Scarabeo 50 DiTech models are already getting close to 120 miles per gallon and use 80 percent less injector oil than traditional two-strokes. Look for Kymco, Honda, Yamaha, and a host of other builders to soon offer direct injection two strokes (assuming regulations permit them).

"I decided to harken back to the Streamline days of the 1940s and design a whimsical '1946 Designer's Dream Scooter for 1950.' It has the proportions and cliche design elements that were so silly but captivating. This is the scooter I wish they had made but didn't, until now."
—Craig Vetter, designer

Fuel Cell Technology

Aprilia is leading the charge into building scooters using hydrogen fuel cell technology. This "zero-emission" power source combines hydrogen from a fuel tank with oxygen from the air to create an electrical current. The only byproduct of this process is pure, clean water. Fuel cells will also appear in all kinds of other products. Hydrogen can be produced anywhere there is a supply of electricity and water. Fuel cells reverse the process, turning hydrogen back into those two components. They are among the most exciting and important technologies in the future of scooters.

Electrics / Hybrids

There is no question that electric vehicle technology will be the big story in the future development of scooters. The only thing holding electrics back is the slow development of battery technology. Rechargeable batteries are still too heavy and they do not hold a charge long enough to allow for commuting on electric scooters. Also, electric scooters are too slow. Only a handful of models have yet been designed that can handle even residential speed limits. In the future, however, science will overcome these problems and electric scooters will dot the landscape.

Honda Motor Company made giant strides in the efficiency of automobiles when they introduced models powered by a combination of gas and electric power. Plans are in place to adapt this same technology to scooters. Small machines powered in this way could easily approach 200 miles per gallon while still providing enough oomph to offer real-world transportation. Honda has already built prototype scooters based on this technology.

EQX

Chapter 2

Scooter Breeds

When it comes to cataloguing scooters, the two prime categories are "modern" and "vintage." Modern includes scooters with automatic transmissions built after traditional imports stopped coming into the U. S. Vintage refers to the classic, typically geared models that populated the planet from the early 1950s through the early 1980s. Scooters built prior to World War II are fittingly referred to as "prewar" models and are not often seen. They reside largely in museum-quality collections, too valuable to be ridden.

Prior to 1998, before the current tidal wave of models from Europe crashed onto North American shores, almost anybody who considered themself a scooter enthusiast rode Vespas or Lambrettas or both. Those were the brands with the appeal, the support network, and the parts availability. Since the dawning of the scooter hobby in North America, these legendary brands have enjoyed a stronghold over the avocation. This is changing to a small degree. It is no longer uncommon to see dozens of modern scooters at any rally. The modern Vespa models—the LX50, LX150, PX150, and GT200—are becoming a very common sight, as are exotic "gray market" imports like the Gilera Runner 180 and the Italjet Dragster. Some enthusiasts have also embraced Kymco, a Taiwanese brand with a very wide range of models and a rock-solid reputation for quality, as well as Aprilia, an Italian assembler known for extremely high-end components and formidable handling. Malaguti, Derbi, Twist-N-Go, Honda, and Yamaha have also released modern marvels worthy of enthusiasm. Bajaj USA and Chicago's Genuine Scooter Company distribute traditional geared scooters, the type once banned in the U. S. These nifty brands are also appearing at rallies in impressive numbers, particularly Genuine's fantastic two-cycle Stella, the only new machine on the market that garners virtually unanimous approval in the rally ranks.

Because the selection of scooters in North America has become so good, it is no longer as tempting to buy an old scooter as it was a few years ago. Scooter seekers once sought vintage models as an alternative to the boxy, plastic Japanese ones that were heavily stocked at dealerships. With the introduction of so many mouthwatering new choices, resale prices on vintage scooters have nosedived. The reason for this is choice. Where scooters are concerned, we now have more choices in North America than at any other time in history. Shopping for scooters, wise consumers will visit at least three stores, where as recently as five years ago they might not have had even one dealer in their town. The challenge has shifted from simply finding a scooter to trying to choose one.

Modern: Just Twist and Go

Lambretta and Vespa imports stopped coming into the North America around 1984, leaving the market vacant of new European scooters. Stricter laws pertaining to emissions and safety standards nixed the larger two-stroke models with manual transmissions and foot-operated rear brakes. Two-stroke engines were, for the most part, limited to 50ccs; shifters were eliminated completely. The Japanese scooter era, the "Scooter Ice Age," had begun. After Honda and Yamaha introduced step-through, twist-and-go models, the very definition of scooters changed in North America from fun, little machines to strictly utilitarian, gas-saving devices. With their cheesy decals, soft, spindly suspensions, and oddly angular styling, these new Japanese machines were not what scooter aficionados expected scooters to be. They offered no collector's appeal at all, nor were they intended to. Rather than being carefully crafted design masterpieces with heavy-gauge steel bodies and sleek lines, Japanese scooters had flimsy plastic bodies and blocky styling. Vintage scooter enthusiasts have long referred to these as "Tupperware" models in reference to their synthetic skins and semi-disposable construction.

With the goal of keeping prices low and volume high, Honda and Yamaha stocked their dealers very heavily in the mid 1980s and sales hit record levels. Dealers could not even keep scooters in stock; as soon as they arrived they were sold. From 1982 to 1998, Hondas and Yamahas were pretty much the only scooters being officially imported into the U. S. The only other choices consumers had were mopeds, mostly from France, Austria, and Italy, which were innocuous enough to escape safety and emissions scrutiny. But most folks preferred not to pedal, so they chose the new breed of twist-and-gos from the Land of the Rising Sun. Japanese scooters from the era featured hand brakes and belt-driven CVTs (constantly variable transmissions) similar to those once offered by Salsbury and Cushman. This arrangement became the standard in the eighties and it remains so today. The modern CVT gives the rider complete control over their scooter's speed with a simple twist of the throttle—no shifting, no fuss, no muss. Hence the term: twist-and-go.

Though the CVT was invented in the 1930s, it was perfected in the 1980s. Some of the scooters designed then are still available new; Japanese manufacturers have only recently begun to replace their aging fleets. The Honda Elite 80, for instance, first appeared in 1986. It has not changed a bit; the boxy plastic scooter of today is identical to the one on the floor when *Back to the Future* was in movie theaters.

The fabulous Aprilia Scarabeo 50 and Kymco People 50

Most modern scooters, however, are updated from those offered in the 1980s. Honda and Yamaha no longer enjoy the type of dominance that allowed them to offer the same old tired scooters, year after year. A huge number of builders and distributors from Europe and Asia have transformed the North American scooter market into a potpourri of intriguing choices. Each new season the ante is raised, and even the two big Japanese firms now introduce exciting new models in an effort to remain atop the sales food chain. Quite suddenly, North America has a selection almost as vast as those in Europe and Asia. In 1997, a consumer seeking to purchase a 50cc scooter could choose from a Honda Spree or Elite or a Yamaha Riva, models that are essentially the same. In 2005, the same consumer could test ride thirty different models from manufacturers from all over the planet.

Honda's 50cc Metropolitan and Metropolitan II have converted many to scootering with their peppy colors and vintage look. This group is from the Scooter Dolls, an all-Metropolitan scooter club.

Fifties: Good Things Come in Small Packages

The overwhelming majority of new scooters sold these days are those with the smallest engines available: 50cc models. Even the very first Vespa had a 98cc engine, almost twice as large in displacement as the best-selling models today. With a piston the size of a shot glass, 50cc engines seem like they are very small, but they get the job done with incredible spunk. Because the little forged chunk of hardened steel within them can travel up and down really, really fast, these motors really, really run. In fact, freed from factory restrictions intended to meet federal regulations, most 50cc models can travel upwards of forty-five miles per hour. That's nearly one mile-per-hour per cubic centimeter of displacement. If cars possessed the same power-to-displacement ratio, a Volkswagen bug would travel roughly twice the speed of sound.

These 50cc models are also popular because many states have special laws pertaining to their registration. In some places, the little ones are treated like bicycles and do not require license plates or a driver's license, and can even park legally at bicycle racks. In other places, little 50cc models can be used by kids as young as fourteen or folks who lost their license to drive other types of vehicles. (The latter has led to 50s being nicknamed "liquor cycles" in some jurisdictions.) Because scooters with 50cc engines are so popular—they make up almost 75 percent of scooter sales—the wide range of available styles can suit almost any taste.

Retros

While North America was suffering through the "Scooter Ice Age," Europeans were exposed to new scooters every year. Factories across Europe spent the period from the mid 1980s to the late 1990s taking scooter technology to the next level, squeezing more power out of little engines, making them cleaner and more efficient. Along with mechanical progress came design progress, as scooter design followed a high-tech course. European customers were (and still are) interested in cutting-edge designs, powerful brakes, fancy suspensions, and sexy angular bodies. Most scooters built in Europe these days still follow this formula, but others are like they were in the good old days when scooters were pretty and they rode on eight- or ten-inch wheels. Nostalgia is a prime motivator for buyers in North America as well as in Japan and other parts of Asia and, to a lesser extent, Europe. To please consumers with an eye for the past, scooter manufacturers offer scooters we call "retros."

These throwbacks usually offer plenty of chrome plating, appealing paint jobs, and flowing lines inspired by the classic European models that started it all. Retro models are not typically the fastest or the best-handling scooters, but they do give their owners what they are looking for: a little nostalgia and a whole lot of attention.

Italjet USA was the first company to bring retro models back into the U. S., the irresistibly cute Velocifero being the first. Built using a traditional Vespa legshield attached to a tubular subframe, the Velocifero offered everything Hondas and Yamahas did not. Unfortunately, the Velocifero lacked the reliability of its Japanese counterparts. The sexy little scooter earned a nightmarish reputation for poor quality; engineers overlooked the fact that the rear tire could tear the gas tank apart when two people rode. Nevertheless, the Velocifero was the scooter that started a revolution. A couple of years after it came out, manufacturers from across Europe followed Italjet's path into the U. S., and sleek, gorgeous scooters like the Malaguti Yesterday, the Aprilia Mojito, and the Vespa ET filled showrooms from coast to coast. By 2002, even Honda and Yamaha had retro models in their lineups. Yamaha's two-cycle Vino and Honda's four-cycle Metropolitan are likely to become all-time bestsellers thanks to their very low price, but like the Japanese scooters of the past, they are generally shunned by "classic scooter" enthusiasts.

Commuter Models

Riding a scooter around an American city like Des Moines is not really all that different from riding around in a place like Paris. One is likely to encounter the same types of inconveniences, potholes, manhole covers, rocks in the roadway, and the like. These realities of everyday commuting long ago led European scooter manufacturers to sixteen-inch wheels. By making the wheels of their scooters larger, manufacturers made them more stable and, often, safer than scooters with smaller wheels. While big wheels make scooters look somewhat like mopeds, they are quite popular with American consumers. Models with sixteen-inch wheels are sometimes called "commuter scooters," ideal for getting to work and back.

Aprilia's Scarabeo models have long defined the commuter category. Their sleek, distinctive lines and

Malaguti Yesterday

superior build quality have set the standard for the other products in the field. The Scarabeo has been available in the U. S. in 50cc and 150cc versions, powered by a thrilling four-valve Rotax engine, as well as a 500cc behemoth propelled by a massive single-cylinder Piaggio plant. These scooters have enjoyed great popularity and were among the most popular in Europe at the dawn of the new millennium. Joining them on American shores were Kymco's excellent People models in 50cc, 150cc, and 250cc versions, the Malaguti Ciak with both 50cc and 150cc engines, and the Piaggio Liberty and Beverly models.

Sport Models

The demand for retro and commuter scooters is pretty simple to understand. The demand for sport scooters, however, is a little tougher to define. Sport scooters are typically among the highest-priced scooters in their categories, sometimes approaching the cost of a motorcycle. They offer high-tech features not seen on retros or commuters such as water-cooling and disc brakes on both the front and rear. The styling is modern, designed to resemble sporting motorcycles, or "crotch rockets," and cater to the European thirst for cutting-edge technology.

San Francisco Scooter Center's Lambretta racer

Aprilia's aggressive SR50

Rather than looking like something from the past, sport scooters look like something from the future. Handling characteristics are given top consideration and, although they produce the same amount of power as other scooters, they are quicker and more nimble. Manufacturers often use the same engines in sport scooters that they do in other models, but they tune them differently and configure the transmissions to offer brisker

The Vespa LX series

acceleration and snappier throttle response. Tires are premium and stick to the pavement during aggressive riding. The result is a modern scooter that delivers as an everyday means of transportation, a Sunday pleasure-tripper, or a full-blown racetrack weapon.

Italjet USA led the charge importing sport scooters to the U. S. with its Dragster and Formula models. These cleverly styled machines featured Minerelli engines and were as fun as they were sexy. Italjet USA also brought in a very limited number of Formula 125 and Dragster 180 models. These super-high-performance two strokes are some of the most popular among scooter collectors. Since they did not meet DOT and EPA requirements, they were available for "off-road use only," but many buyers found a way to attain license plates nonetheless.

Aprilia was the next major brand to import sport models. The Italian firm released its formidable SR50 model in the continent in 2001 to rave reviews. The powerful scooter featured disc brakes front and rear and was liquid-cooled, allowing for greater modification. In its third year of availability the SR50 got updated with Aprilia's advanced "direct injection" or "Di-Tech" system. This feature rendered the SR50 the cleanest, most efficient scooter on the market with fuel mileage exceeding 100 miles per gallon and oil consumption reduced by nearly 80 percent. Next, such manufacturers as Malaguti and Derbi began shipping sport scooters to North America and the genre broke wide open. Buyers can now select from a host of outstanding offerings in the category, all with powerful brakes and water-cooling. Most sport models are 50cc machines, but, as their popularity grows, so does the selection of styles and sizes.

Mid-Sized Scooters

The majority of new scooters sold in North America are 50cc models with top speeds below forty miles per hour. While smaller machines may be tops in sales, mid-sized scooters might offer the best combination of features for most riders. Because they are not much larger than 50s, but offer significantly more power, they are the masters of the boulevard. Ranging between 125 and 200ccs, mid-sized scooters offer low mass and rapid acceleration. They are usually highway capable and allow for light-duty touring on back roads and two-lane highways. Top speeds vary among the models but most are capable of maintaining speeds between fifty-five and seventy miles per hour.

Mid-sized scooters feature four-cycle engines, either air- or liquid-cooled, and usually have powerful front and rear disc brakes. They weigh on average around 250 pounds, a comfortable weight for most riders. Unlike 50cc models, which run at a very high RPM even on residential streets, mid-sizers have extra passing power on tap when they are ridden in the city. Mid-range acceleration is a top priority for manufacturers.

Aprilia's Scarabeo 150

North American consumers waited longer for mid-range models than they did for other sizes of scooters. As the new scooter market developed, manufacturers rushed 50cc and even 250cc models into dealerships but lagged on with the mid-sized category to the frustration of dealers. Part of the reason for this was that they were still "testing the waters" to see how well people would respond to scooters. Also, the machines spent more time in DOT and EPA testing because of stricter guidelines on larger engines. Vespa won the race to put a legal mid-sizer into U. S. dealerships with the amazingly popular ET4 150, the scooter Americans were waiting for. Vespa "boutiques" all over the United States saw the ET4s sell just about as quickly as they could unpack them. Vespa replaced the ET series with the LX in mid 2005, showing some improvements and stylistic refinement influenced from the larger GT200.

Also among the first mid-sizers to make it to market was the awesome Malaguti F-18 Warrior, the larger brother of the 50cc F-15 Firefox. The radically racy Warrior included a stone-reliable Kymco built engine, liquid-cooling, and front and rear disc brakes. For consumers and dealers alike, the Warrior represented the best of both worlds in terms of size and power. Roughly one summer after the release of the Warrior, Derbi introduced its Boulevard 150. Powered by the very popular Piaggio Leader 150 engine, the Boulevard is smooth, fast, and dependable, making it even more popular than the Warrior.

While most scooters in this class are styled to resemble sporty racing scooters, the Vespa GT200, released in 2004, like the ET4 150 before it, resembles the Vespas of yesteryear. The GT200 is the culmination of decades of research and development and is the pride of the mid-sized category. With a steel monocoque chassis, a phenomenal engine, and looks that kill, the scoot somehow justifies its enormous price tag, far and away the highest in the category. But then, most buyers will tell you that the Vespa GT200 is in a class all by itself.

The Minnesota Maxiscooter Club love their scooters big and comfortable

Piaggio's BV500

Maxi Scooters

No other category of scooters induces disbelieve like maxis. People often do not realize that machines like this exist. They represent a brand new product category in North America. Popular in Europe for their freeway capabilities, maxi scooters are designed for touring outside of the city. Ranging in displacement from 150 to 600ccs, maxis are noted for their fully paneled bodies, ample storage areas, and substantial protection from the elements.

Maxis are similar in concept to the great touring motorcycle, the Honda Goldwing. Creature comforts are numerous and the rider sits comfortably inside a bubble of still air. Many models include such features as cellular phone jacks and built-in communication systems. All offer plenty of power on tap, so much power that potential buyers sometimes ask, "Why shouldn't I just get a motorcycle?"

The answer is simple: Although maxi scooters are pricey, costing upwards of $4,000, they're cheaper than all but the most basic motorcycles, yet they offer premium features. The maintenance cost is also lower than that of a big bike due to the simplicity of the engines.

The most popular and best-selling maxi of all time is Honda's homely Helix, which debuted in 1986 and resembled a vehicle from *The Jetsons*. Sales did not meet Honda's expectations, so the company dropped it from the lineup after just one year. When the supply of new Helixes finally dried up in the early 1990s, Honda re-released the beast as a 1992 model with no changes from the earlier version. A dozen years later, Honda began offering the Helix once again as a 2004 model, eighteen years after it was first introduced. The Helix was undoubtedly a pioneering product in the scooter industry, regardless of its unusual looks.

Maxis began to really make their mark early in this century. Taiwan's Kymco demonstrated its 150cc and 250cc Bet & Win models at the Vespa Club of America's 2002 Amerivespa Rally in Knoxville, Tennessee. The sleek and powerful Kymco would be the first in a long line of new touring scooters to be released over the next two years. That same year Honda introduced their Reflex 250, a pricier, more modern alternative to the mighty Helix. Rumors flew that Piaggio was soon to introduce its X9, a 500cc machine that was creating a buzz in Europe. The much ballyhooed tourer was slow to make it to U. S. shores, however, and Aprilia quickly introduced their own machine propelled by the same engine as the X9: the massive Atlantic 500. Piaggio did release the X9 Evolution 500, followed with the BV200 and BV500 models.

Soon to follow the Atlantic 500 was Honda's even-bigger Silverwing 600 and a pair of enormous Suzukis. Suzuki, one of the Japanese "Big Four" motorcycle builders, had never been known for scooters in North America, but the company's nicely appointed maxis—the Burgman 400 and 650—were extremely

Genuine Motor Company's Stella

Bajaj Chetak

Vespa PX150

Traditional models are among the best sellers in today's scooter market. Their manual shifiting, sturdy steel bodies, and ten-inch wheels may not be cutting-edge but the models provide a familiarty that more modern designs simply can't. Genuine's Stella and Bajaj's Chetak are built in Asia, while Vespa's PX150 is once again being produced in Italy.

popular in Europe and Asia prior to their release here. Since the firm decided to bring the machines to North America, they have developed an extremely loyal following. As the twenty-first century gets older, the maxi category continues to expand by leaps and bounds.

Vintage: They Just Don't Make 'Em Like They Used To

When people buy a modern twist-and-go scooter, they do so for mostly logical reasons. Maybe they need an alternative vehicle to complement their car. Maybe they have short commutes and want to take advantage of free parking and low operating costs. Maybe they just need it for the back of the Winnebago or to tool around the fairgrounds during horse shows.

Whatever their reason for buying, most modern scooter owners have true enthusiasm for their machines. But owning a modern scooter is typically not a passion, or even a hobby. Vintage scooters are a very, very different story. While they offer most of the same economic benefits of their newer counterparts, they are

generally purchased as hobbyist vehicles rather than as mere tools of transportation. Vintage scooters are often a reflection of their rider's personality, a reward for their dedication. Vespas, Lambrettas, and other antique scoots can be as much a part of people's lifestyles as they are part of their lives.

Vintage scooter enthusiasts are motivated to own, collect, and ride their classic machines by their beauty, their functionality, and their cool factor. It doesn't hurt that the perceived difficulty associated with their upkeep keeps old scoots unique and just outside the grasp of the general public. Because there is an almost endless list of ways to personalize and customize an antique scooter, it is not difficult to build one that reflects any personality or taste. Some are decorated with chrome crash bars and tons of mirrors. Others are hopped up with giant carburetors. Still others are austere, completely original and unaltered. It's all up to the owner. An old scooter is a personal canvas, a tattoo that can be ridden in rush hour traffic.

Picking the Perfect Tattoo

When scooterists cruise down the boulevard aboard a vintage machine, they know that not just anybody can enjoy this by simply opening a checkbook at the local dealership. Owning a vintage scoot involves a higher level of commitment, effort, and technical familiarity than owning a modern. The commitment starts with the search for a scooter. While the availability of vintage scooters has improved, the affordable "garage find" is increasingly elusive. People who might have been storing them for a decade or two have learned of the increasing value of scooters either through word of mouth or by trolling the Internet. As a result, scooter seekers are no longer able to find antique models at garage sales, swap meets, and abandoned farms. Those highly regarded types of finds are now priced at whatever the market will bear, not whatever the buyer will offer.

Unrestored: 1970 Lambretta GP150
Owner: Ryan Basile

Ryan Basile (a self-described "Lambretta Guy") discovered his beautiful, orange-colored GP in 2001. It was hiding in the background of a photograph he received via e-mail. Ryan says, "A guy in the U. K. sent me photos of some rare European scooters he was selling, and the GP was just sitting in the background. He did not even intend for me to see it."

This thirty-something-year-old unrestored scooter arrived in wonderful shape, complete with the owner's manual and factory tool kit, both items that elevate the GP's stature within the unrestored genre.

Like most machines that have wandered the earth as long as this one, it has some small flaws. Touch-up paint was once used to repair scratches in the headset, although it appears to have been original factory paint. Ryan has also added some appealing "period" accessories, which, because they are not reproductions, do not detract from the originality. The sporty shield was a must-have item in the early seventies.

Because of its excellent condition, its scarcity, and its unusual orange color, this is one collectable Lambretta any hobbyist would love to own.

Unrestored: "Decadron" 1960 Lambretta Li150
Owner: Colin Shattuck

I first learned about "Decadron" in 1999 when an old fellah from Cheyenne, Wyoming, stopped at the scooter shop one day on his way through Denver. Charles told me he had a Lambretta in his basement, probably a '58. I assumed it was another rust bucket old Lambretta LD.

A year later Charles came back to the shop with photos of his prize, which turned out to be an Li150 Series 2 model. It was beautiful. It had received such good care through the decades that its original tires were still supple, its rubber seals and trim parts immaculate, its accessory windscreen and bonnet un-faded. It looked as though it had been dipped in candy—red and chalky-white, shiny and immaculate. After a long negotiation period, Charles and I worked out a deal with a fair price and a lot of respect.

Being that Decadron was an all original, I confess that, despite running her almost 2,000 miles, I have never done more than change her oil. Decadron is an anomaly; a nearly immaculate decades-old Lambretta that is just as smooth and reliable as the day she was born. She is named after the funky pentagonal key tag that was attached to her original keys on the day I got her. This scooter is like a vacation back in time.

Once a fitting scooter is found and purchased, it often needs to be revived before it can be ridden. Parts have to be ordered, buddies called, cobwebs cleared, engines rebuilt, paint jobs funded. Sometimes a scooter shop is involved, other times owners rely on service manuals and mail-order houses like Chicago's *Scooterworks USA*.

When the work is done and a scoot is ready to go, it has to be registered. This can be a frustrating step in the process. Often it means title searches, surety bonds, paperwork, and patience. Because vintage scooters are, well, vintage, it is likely that the original title was misplaced long ago. Depending on the particular state in which the scooter is to be plated, replacing the title can be very difficult and very expensive. Title services can be a big help.

Once a vintage scooter is repaired and registered, it must be maintained. There are cables to be changed and adjusted; there are bearings, bushings, and brake pads; there are fuel delivery tics, air intake issues, and electrical problems. Yes, that ride down the boulevard on a classic scooter, all eyes on you, that is something earned, something deserved, something rare. If it were easy to keep a vintage scooter on the road, it would not be nearly as fun.

It's tough for a lot of folks to understand that; especially individuals who have sunk several thousand dollars into a machine that they rarely get to enjoy. Some people were just not meant to own them, let alone maintain them, which is an art in itself.

Old scooters are like snowflakes, no two are exactly alike. Even two seemingly identical machines, built in the same factory on the same day, can feel as different as a Porsche Boxster and an old VW bus. Picture two 1978 Vespa P200s, each with a fading silver factory paint job, the exact same mileage on the odometer, even the same torn seat cover. As similar as these two scooters are, there is an excellent chance they have entirely different personalities. The maintenance they have gotten (or not gotten) over the years, along with the major repairs made, the parts replaced, and the modifications performed, give each and every vintage scooter its own distinctive feel and character. To its owner, a particular scooter might feel like the best one ever built: fast, smooth, and safe. To a friend, the same scooter might seem like a deathtrap, barely worthy of being ridden. Quality is in the eye of the beholder. Like a car with a tricky clutch, a vintage scooter, like the people who own them, has its quirks.

There are just about as many styles of vintage scooters as there are types of people who ride them: restorations, racers, even low riders. Scooters have been built to resemble fighter aircraft, chopped Harleys, John Deere tractors, and pro-street dragsters. They have been cut, chopped, lowered, lengthened, hacked, and welded. They have been painted every color imaginable, flamed out, and given wild splash graphics. They have been dedicated to rock and roll bands and decorated like bottles of beer.

Vintage scooters mean something different to each and every person who owns one, which is their beauty. Stand back six feet; let your mind run wild. What would you do to change it? Would you leave it just the way it is? What color should it be? Should it have whitewall tires? Luggage racks? This is the process almost every scooterist goes through the day a new scooter first comes home. People keep chairs in their garages just so that they can sit there, stare, and imagine what that scooter is going to look like someday.

As the hobby has grown in North America, so has the quality and creativity of scooter builders. To many scooterists, the machines are prized possessions, more valuable than their houses and cars, more precious than wedding rings and baby photos. They have invested their souls into the things, forging a relationship with a conglomeration of steel, aluminum, and rubber. It's a difficult bond to break.

Unrestored: 1955 Vespa 150 VL-1
Owner: Mike McWilliams

Ever since the early 1950s, Shrine Temples throughout the United States have chosen Vespa scooters as parade vehicles along with Jeep, Go-Cart, Cushman, and Goldwing. Each troop is allowed to determine their vehicle of choice, and then each member must decorate his in precisely the same way. Shrine Vespas are some of the most sought after by collectors because of the Americana factor.

This stunning Vespa 150 was once owned by Estel "Tommy" Thompson, a member of the Al Kaly Shrine Temple and part of the Legion of Honor organization that participated in Shriner parades in Kansas, Arkansas, Nevada, Missouri, New Mexico, and Colorado.

Michael McWilliams, long time president of the Vespa Club of America, picked up this wonderfully preserved shrine scooter in 1993. He purchased it from a representative of Thompson's estate for a paltry $100.

"It needed an engine rebuild," Mike explains. Still, he stole it.

Ironically, Mike found another scooter identical to this one a couple of years later. "All of the accessories had been removed, but the mounting holes were still there so you could see where they had been."

The second shrine scoot, which Mike later sold, had a vehicle identification number only ten digits off from the one he kept. "They were sister scooters," Mike explains. "In retrospect, I wish I had not sold the other one."

Mild Custom: "Vader" GS160
Owner: Paul Kavinaugh

"Everything just came together when they built this bike," claims Paul Kavinaugh. "Just look at the design. The GS160 is the quintessential Vespa in my mind."

Paul knows a thing or two about design. As a lead animator for Industrial Light and Magic, a division of Lucasfilm Entertainment, Paul has a keenly developed sense of style.

He continues, "I had a P200 when I first moved to San Francisco. I rode it into the city every day from Sausalito. I loved it but I knew I had to have a GS160. The problem was that they cost too much in Northern California."

Paul found an alternative to shelling out the $6,000 to $7,000 GS's were selling for in his region. He purchased this black beauty from a German reseller; had it packed into a box and loaded onto an airplane.

Paul knew that because he was buying a scooter from overseas, he would need to make improvements. "I immediately replaced several things," Paul says. "My friend Josh at Bulletproof Scooters in S. F. helped me replace the coil, the top end, and the cruciform."

How did Paul feel about disassembling his brand new GS? "It was scary," he admits. But the work paid off. Paul's stunning GS has proven to be a reliable steed.

Unrestored Originals

The most cherished of all vintage scooters are those that have survived the rigors of weather, time, storage, and use with the least amount of alteration or damage. As with most collectible objects, restored scooters are not as valuable as those that are beautiful yet unrestored. Scooters that have retained their factory finishes and original "feel" reflect history in their shiny paint. Having ridden hundreds of scooters in various states of repair, it thrills me to have an all-original scooter fire up on the first kick, slide smoothly into gear, and gently pull away. It does not matter if that scooter was built in 1958 or 1977, factory condition is the best condition of all.

Some people buy scooters and immediately begin to never, ever use them. They might be frightened because of an incident on the way home from the dealership. They might just find scooters uncomfortable. These folks have a difficult time swapping the security of their automobiles for the freedom of the scooter, and the relative danger that comes with it. Because of non-use, these scooters sometimes go on to become "garage finds."

Garage finds are virtually non-existent in parts of the world where scooters are vital daily transport, like Italy, India, and Vietnam. In those places, and in most of the world, scooters endure extreme use. They carry more weight than they are designed to (entire families pile on), the roads are rough (assuming there are roads), and the maintenance is minimal.

It is even difficult to locate garage finds in the home country of Piaggio and Innocenti. A thirty-year-old Vespa still in use anywhere in Italy is likely to have upwards of 60,000 kilometers on it. Though the Italians do an excellent job of maintaining scooters, they wear out Vespas like Americans wear out Nikes. In contrast, many American scooter buyers from the 1950s through the 1980s were very gentle on them. A significant number of classic scooters made it to North America and, while that number pales by comparison to the number sold in other parts of the world, the percentage that have survived in beautiful condition is quite high.

The most appealing thing about original machines is the fact that they cannot ever be replaced, only simulated. A restoration is an attempt to make a scooter precisely what it once was, but is most often a futile one. A good restorer can match the proper paint and primer codes, plate all of the hardware correctly and assemble a scooter with a craftsman's eye for detail. They cannot, however, re-create factory flaws. Repainted scooters never look quite right,

Mild Custom: "Artimus Sprint"
Owner: Natalie Lachance

"My favorite thing about my scooter is that it makes people happy," claims midwife-in-training Natalie Lachance of Ottawa, Canada. A recent convert to vintage machines, Natalie once rode a Yamaha Beluga; an exceedingly angular automatic built during the great "Scooter Ice Age."

Natalie built this eye-popping mild custom without breaking her piggy bank. "It started as a mid-seventies style Sprint with a bent frame," Natalie explains. "I found a '76 frame for free and we bought the '66 trapezoid headset on e-Bay. The floor boards were bad in the free frame so we welded in new ones from ScootRS."

This conglomeration of parts all came together as a example of how a vintage scooter can be both affordable and attractive at the same time when a scooterist offers up the elbow grease.

Inspired by the alternative clothing and accessory line, "Emily the Strange," Natalie painted this scooter herself with a little help from her fellah, Glenn. "We basically spray bombed it," she says, admitting to having used hardware-store spray paint. "The fun part was hand painting the graphics."

The dizzying red swirls and cunning black kitty makes this scooter, nicknamed "Artimus," stand out in a pack. The seat cover, which was custom made by a clothing designer, also garners compliments.

Mild Custom: "Orange Crate" Vespa GS160
Owner: Dean Wright

"I have always loved the Schwinn Sting Rays," states Dean Wright. "I like those shiny fruit colors they came in."

Dean, who once built custom bicycles for a living, is a production engineer for John Deere and represents the Midwest with this stunning "Orange Crate" Vespa GS160. A do-it-yourselfer, Dean built his Vespa from the ground-up.

"It was pretty beat," Dean recalls. "The back end was all smashed in and the fender was bent up. Also, the cowl louvers were all busted out."

Luckily, Dean has experience in metal fabrication. He did all of his own bodywork. "I never could have afforded to pay someone else to do it," he says. Dean also painted the scooter himself using "House of Kolors" metallic paint and inspiration from a famous English scooter dealer.

Dean explains, "I based the paint scheme on the Eddie Grimstead Hurricane scooters from the 1960s." Grimstead was famous for his customized dealer specials, which were a hit with the mod movement.

not even those done by the best restorers in the business. Things that might have been hastily installed in the factory are given careful consideration in the workshop. A fellow in the factory might have installed a decal slightly cockeyed or a seat cover askew, but a restorer seeks perfection. They measure twice and cut once, noticing every flaw.

Part of what makes an unrestored scooter special is its tiny flaws, earned through decades of either faithful service or irresponsible storage, most of which the owner could point out blindfolded. Garage dings, little scratches, and gasoline stains are each chapters in its story.

Restorations

The word "restoration," like the word "love," is used incorrectly more often than not. According to *Miriam Webster's Collegiate Dictionary,* to restore is "to bring back to or put back into a former or original state." Too often scooters are represented as being restored but are actually refurbished. What is the difference? A restored machine has been put back the way it was, down to the last detail. A refurbished one might have different paint or a more modern engine; more like a mild custom than a true restoration.

The words "fully restored" appear frequently where scooters are found for sale. In all actuality, fewer than 5 percent of scooters advertised as such really are. Restorations are expensive. Aftermarket parts are unacceptable. Only the rare, original stuff will do. A professional restoration will almost always cost more than the scooter will be worth on the open market. True restorations are done for the love of the motorscooter, not for gain.

The finest restored scooters in the nation are the handiwork of private collectors, not paid professionals. Restoring is a meticulous hobby enjoyed by people who typically don't trust others to do the job right.

Mild Customs

Some people find restoration boring or far too tedious. They prefer to pick out fun colors, buy custom seat covers, and hop up their engines. Mild customs are scooters that are not technically "restored," but done up in a manner that reflects the personalities of their owners. They might be bone-stock with flashy paint. They might have a modern engine. They might have badges and trim from a 1950s Chevy. They might just be entirely covered in racing decals. Mild customs are the most common type of vintage scooter and the degree of modification varies widely. Most scooterists personalize their scooters in some way, no matter how small, in order to set them apart.

Radical Customs

What the radical mind can conceive the radical mind can create. Some scooters bear virtually no resemblance to their former selves thanks to the people who create rolling works of art based on their own visions and illusions. A radical custom is a scooter with only a handful of unaltered pieces, one that boldly defies the initial architect's intentions with upgraded brakes and suspensions, with screaming hot-rod motors, and expensive exhaust pipes. Most importantly their frames and bodies have been chopped, bent, cut, and reshaped beyond recognition. The category is a free-for-all of artistic and mechanical expression.

To compare radical customs to cars, these

Radical Custom: "The Batbike"
Owner: Paul Italiano

"Holy bazookas, Batman! Get a look at that crazy custom scooter."

Paul Italiano offers this tribute to his favorite comic book hero. Drenched in thick, gloss-black paint and decorated with incredible airbrushed murals, the Batbike turns heads everywhere it goes.

But its beauty is not just skin deep. This is one of the most radical custom scooters anywhere. There are only two pieces on the Batbike that are original: the right control lever and the left control lever. Everything else has been hacked, sliced, bent, or replaced. Bat Bike began a second life when its corroded, forgotten old frame was first pulled out of the weeds. It was disassembled and sandblasted. Paul did not even have an engine for it then … just a vision.

Batty is filled with the best performance equipment money can buy. Paul spared no expense in making sure that this comic book hot rod can really cook. He purchased an electric start P200E engine brand new out of the box and immediately ripped it apart. He added a Malossi brand 210cc cylinder kit, an aftermarket carburetor, and a massive expansion-chamber style exhaust. In order to put more meat on the pavement, Paul added a wide, rear-wheel conversion kit made by a German company. And he did not overlook braking and suspension. The shocks are adjustable Italian Bitubo units and a powerful hydraulic front brake was added in case Batty ever needs to stop on a dime.

How much did all of this cost? Don't ask Paul. He tried not to add up the receipts.

Radical Custom: "Liverpool Football"
Owner: Damir Gusic

"I get tired of my scooter," admits Damir Gusic, an employee at Motoretta Scooters in downtown Toronto. "Rather than sell it, I just change it."

Unlike most scooters featured in this book, which were old, banged-up projects before restoration, this "Frankenbretta" was already looking and running good before Damir tore it down. "I got tired of all the accessories. I had to worry about tightening them all the time and keeping them from getting rusty. Besides, my scooter was too heavy; it was like riding an SUV."

In its current form, Damir's Lambretta is a tribute to the Liverpool Football Club. The phrase "This is Anfield," on the front fender refers to the squad's home stadium. The unusual green color was borrowed from Porsche. "They only used it in 1972," Damir explains. The shade reminds him of a table soccer game he owns. "The grass pictured on the game's box is almost that exact color green, and the game is a Liverpool Football Edition. I guess that's why I made the association."

While the scooter appears to have a race-inspired appearance, Damir did not originally set out to build a racer. "Every piece on the scooter was chosen with reliability and ride quality in mind," he states. The fact that it wound up being a hot rod was pure accident.

are the "pro-street" hot rods of the circuit. Built for speed and for getting noticed, these scooters blur the line between cute and downright nasty. A true radical custom costs a whole lot of money and requires a whole lot of time. A good working relationship with a machine shop doesn't hurt, either.

Racers

Built for a single competitive purpose, racers are light, nimble, and minimalist. No piece is retained that does not directly contribute to shortening the scooter's prospective lap times. Every possible ounce is shaved and the original appearance of a scooter is decimated in the name of raw speed. There are various classes: stock classes allow for only mechanically unaltered rigs, while "unlimited" classes allow just about any scooter under the sun.

The very term "racing scooter" might seem like a bit of an oxymoron. Scooters are an entertaining form of light-duty transport, not performance-oriented riding equipment. On the right type of racing surface, however, they deliver a spirited and affordable form of motor sport action. Even racing go-carts cost more than racing scooters.

Choppers

Chopper scooters are just like chopper motorcycles—only funnier. Most people cannot witness a chopper scooter and not laugh. They mock big choppers, waving a finger in their faces and telling them not to take themselves so seriously. Common in the U. K. and increasingly popular in North America, choppers are the wild children of the scooter genre. Stretched and lengthened often to ridiculous degrees, these scooters feature ape-hanger bars, low-mounted seats, and a minimal amount of body work.

Ratbikes

Not everybody prefers shiny paint. Some scooters are downright ugly. Held together with chicken wire and covered in bumper stickers and sticky Krylon, ratbikes are the unclean vermin of the scooter gutter. Their owners mount potato cannons and bowling trophies and bizarre items never intended for bolting to a scooter. These scoots are smoky and smelly and unappealing—intentionally.

Ratbikes pay homage to the days when scooters were plentiful but parts and mechanics were not. Farmers and other rural individuals invented "bodges," unconventional repairs that allowed them to eke more out of their scooters in the absence of needed resources. They ran trailer lighting, rigged external shifters, and replaced brake cables with rebar. They slathered on thick coats of house paint, added tractor seats and makeshift trailer hitches. They cut parts of the scooter off and tore giant holes to make maintenance points more easily accessible.

Inventiveness abounds in the ratbike category. Builders are as clever, if not as technically savvy,

Radical Custom: "Black Hornet" P200
Owner: Adam Baker

Adam Baker's strikingly simple and undeniably menacing venomous black P200E was built for speed. Adam, on the other hand, admits that he is not. "To be honest, I am not really sure how fast it really goes. At seventy-five it starts to get crazy and I back off of it." That is not to say he doesn't ride it plenty hard.

Adam bought the Hornet in 2001 from a customer of his. "We had worked closely with him building the scooter and I really liked what he was doing. When he needed some cash to move to California, I snapped it up."

Adam loved the way the customized P200 looked. "To me it is like a Ferrari. It looks like a woman yet it is not effeminate. It's very masculine, aggressive." From its dropped SIP handlebars to its fattened rear wheel, this scooter means business. The hot rod appearance is highlighted by a chrome front fork, a cut-down T-5 front fender, fiberglass SIP racing seat, and a fully hydraulic Grimeca disc brake assembly.

"Vespa means wasp, and I think a hornet is a kind of wasp, so the name fits," Adam says. The name also fits because this scooter flies. Its red, powder-coated engine case conceals a string of aftermarket expenditures including a Malossi 210cc nicasil-lined cylinder, a 32mm Mikuni carburetor, an electric fuel pump, a lightened flywheel, and a high-performance clutch. Needless to say, Adam rides the Hornet regularly.

Michael Anhalt, a broadcast television engineer by trade, retired from racing scooters with California's ASRA in 1996 after just two seasons of action. He had done well; ranking thirteenth among the club's many members, but grew weary from racing's demands.

Rather than sending his scooter out to pasture, his little Lamb got treated to a shiny new coat of black and the heart of a lion. Its days on the track were over but the time had come to terrorize the streets of Bakersfield, California.

"On Da Lamb" is an example of what is known as a *Frankenbretta;* a Lambretta assembled from several salvage bike parts. This one has the frame of an Li125; the tail lamp and horn cast from an SX200; and the headset, side panels, and leg shields from a GP model.

In order to improve the projects' handling, Michael shortened the distance between the headset and the front wheel by chopping the fork tube on the frame, cutting down the fork and shrinking the leg shield and horn cast. The resulting 'squattiness,' along with its thick, black paint, gives On Da Lamb a nasty, menacing look.

No hot rod worth its salt flats runs around with a wimpy little engine. That is why Michael's Lambretta breathes fire through a huge 35mm Mikuni carburetor. "I have had the needle pegged at ninety-five miles per hour," boasts Michael.

as those who build customs and racers. They wear the title "ratbike rider" on their sleeves and relish the detractive comments others make everywhere they go.

Mod Scooters

When accessories such as mirrors and air horns and flags almost entirely obscure a scooter, a mod must be near. A mod scooter is the epitome of form over function, featuring as many accessories as possible. Mods are dedicated to appearance. Their style of dress is tidy and retro, and love to show off their scooters.

The legendary rock opera, *Quadrophenia*, the story of violent clashes between scootering mods and motorcycling rockers, exposed the North Americans to British mod culture. Soon after the film's release in 1979, scooter kids embraced the lifestyle and built the first mod scooters in North America. The style never died.

The first critical component in assembling a mod scooter is the chromed steel "crash work" that provides mounting points for other accessories. Some believe that the original purpose for all of these accoutrements was to make scooters wider, bigger, more imposing in

traffic so that they offered greater safety. Others think mod scooters owe their existence to simple one-upmanship. Regardless, the style is one of the most distinctive and easily recognizable in all of mechanized travel.

Oddballs

Not every old scooter is a Vespa or Lambretta. Heinkels and Zundapps, Maicos and Ducatis, Fuji Rabbits and Mitsubishi Silver Pigeons—these are all scooters that played important, if limited, roles in the long and storied history of scooter manufacturing. Today these classics still make the show on occasion. Collectors bored with the standard fare often take pride in maintaining rare and unusual scooters. With parts extremely scarce and technical information possessed only by hardcore enthusiasts in unusual places, a ride on an oddball is an even more elusive experience than one on an unrestored original Vespa or Lambretta. To own one means to be granted instant membership into an exclusive club.

Riding unusual scooters, particularly those built in the 1950s, sheds some light on the reasons why Vespa and Lambretta hit such huge success. Many oddball scooters, particularly those from Italy, were "me too"

Radical Custom: "Treasure Hunt" Lambretta
Owner: Nyle Schafhauser

Thirty-six year old Nyle Schafhauser loves Hot Wheels.

Named "Treasure Hunt" after Mattel's limited edition die-cast cars, this awesome Lambretta custom boasts an almost unimaginable level of detail and refinement.

Looking at Treasure Hunt one sees evidence that Nyle makes his living as a designer. One assumes the Hot Wheels logos, which appear several times, are vinyl decals. That's incorrect; all of them are painted on. "I designed the whole thing on a computer," Nyle says. "I put in all of the dimensions and mocked it up electronically before I built it."

Perhaps the coolest treatment Nyle gave the scooter is the trick inside the speedometer shell. Rather than a functioning unit, Nyle shoehorned a tiny mock up of an automobile interior, complete with a steering wheel, three-point harnesses, and bucket seats underneath the clear plastic lens. Cooler still is the scoot's incredibly rare Wal-Phillips brand fuel injector, an ultra slick performance accessory from the 1960s. Very, very few Lambrettas have this.

The beauty of this scooter is in the details. Each and every hanger, fastener, and tiny piece of hardware was given consideration in creating this awesome custom. As it is written on the side panels, this scooter truly is "#1 of 1."

Radical Custom: "U. S. Customs" Racer P200
Owner: Josh Snow

Sometimes simple projects can get a little out of hand. "I built this scooter specifically to sell," confesses well-known custom paint specialist Josh Snow of San Francisco. "I never meant for it to go this far."

By "this far," Josh is referring to the extensive body modifications and performance engineering that went into building this beautiful radical custom. Josh decided to make one seemingly simple modification, which ended up changing the scope of the project entirely.

He chose to use the long, sloping horn cast from a rare eighties Vespa to give the front of the scooter a distinctive look. The Bajaj cowls, which he eventually managed to make fit, lent the rear a sloping, angular appearance. He added a fiberglass racing seat base, complete with an access hole for the slick quick-fill racing gas cap he installed.

He custom cut sections of stainless steel "tread brite" material to replace the floor rails and center mat and to highlight the fork link, fuel lever, and the frame curve just forward of the seat. He then chose a modified front fender, German SIP brand dropped handlebars, and a smooth little aftermarket tail lamp that he integrated into the rear body. To top the cosmetic cake, Josh applied a phenomenal one-of-a-kind paint job featuring the colors of the Italian flag.

"It was already fast when I bought it," Josh admits. He made it faster. The engine in this racy custom includes a list of performance items that reads like a catalog. All of this adds up to a scooter with super-snappy throttle response and wicked acceleration.

products rushed to market too quickly in the frenzied buying atmosphere of post-war Europe. Others, like those made in Germany, were extraordinarily well engineered, but they are the exception to the rule.

Vintage/Modern Hybrids

A growing trend in custom scooters is retrofitting vintage models with modern engines and CVT transmissions. The end result is a custom scooter that offers the best in looks, performance, and ease of maintenance. In the automotive world, hot-rodders have always done this, shoehorning giant Cadillac engines into antique roadsters and installing high-performance braking and suspension components.

For hybrid scooters, the engine of choice is the lump found in the legendary Gilera Runner 180. With power output rivaling motocross race bikes, the Runner's rock easily propels the heaviest old scooters

Sidecars make hauling your pup a little easier

to speeds the factory never dreamed of. The plant from the Vespa ET4 150 is also common in vintage/modern scoots.

Sidecar Rigs

When the family scooter is the family car where do the kids sit? How about the family dog? In the sidecar, of course. Sidecars instantly make a scooter built for two into a sweet ride for three. Some have suspension, others do not. Some mount to the left side of the scooter, others to the right. No matter what, they are difficult to manage. When cornering, a rider must either lift the sidecar up into the air or push it down hard into the ground. It is not for the inexperienced.

Sidecars also offer an entire new surface to paint and decorate. For this reason, they are often the canvas for beautiful paint jobs and cool custom touches.

This gorgeous TV model Lambretta belongs to Richard Guile

Mod Bike: "Chrome Dome" Vespa 150
Owner: Dennis Memmott

The best things in life are free. Just ask Dennis Memmott of Ready, Steady, Go! Clothing in Vancouver, British Columbia. Dennis acquired this mid sixties classic for the price of a deep breath: free.

"I was living with a family in Seattle who gave it to me as kind of a joke," says Dennis. Rusted, weather beaten, and covered in dents, the three speed 150 was left in Dennis' bedroom as a gag. "I immediately started visualizing paint schemes," says Dennis. "I moved it to my basement and began pulling it apart."

This cool mod Vespa features twenty-two stadium style mirrors on eight chrome stems all attached to a chrome front carrier. Extra lighting duties are handled by a pair of extremely rare Yankee Clipper brand spot lamps made in the mid 1940s, as well as a couple of smaller units called "Masons."

To assist in the propulsion of such a burdensome grouping of bits, Dennis replaced the haggard original engine with a late 1970s P125 unit. In addition to giving the scooter a much needed fourth gear, the more modern engine bumped up the over all reliability of the scooter.

While Big People's Scooters in Seattle rebuilt the engine, Dennis performed his own body and paintwork, transforming what was once given to him as a joke into the great looking scooter you see here.

Ratbike: "Thunderturd"
Owner: Dave McCabe

Few ratbikes are built with historical inspiration. Yet Dave McCabe looked to the French military's failed effort to utilize the Vespa as an air deployable weapon in the late 1950s for his muse. The machine they designed, called the T. A. P. (Troupes Aero Portees), featured a large and powerful bazooka mounted across the seat top and through a large hole in the leg shield.

"The inspiration for the 'Turd came largely from the fact that the leg shield was already mangled in such a way that cutting out the hole for the cannon would not be too sacreligious," claims Dave. "When I got the bike it was just a frame," he continues, "leaning up against a house in my neighborhood. I asked the guy who lived there and he said I could have it for free."

Dave plucked a P125 engine from his stash of scooter bits and began assembling his beloved 'Turd. "A neighbor of mine had eighties Thunderbirds sitting in his yard," Dave offers. "He agreed to let me snag the emblems off of them. By rearranging the letters I was able to spell out 'Thunderturd.'"

Dave believes that his ratbike is a rolling social commentary. "I hope that the bike sheds light on the foolishness of war," he told *Scoot! Quarterly* magazine. The cannon mounted on the 'Turd is a very effective, butane powered potato launcher. "It will shoot a PBR can the length of two football fields," Dave boasts.

Step by Step: A Restoration Guide

Take it Apart …

First, disassemble the scooter entirely. Care must be taken to ensure that all parts are accounted for and stored in a manner that will make them easy to access later. The disassembly process must be orderly or every following step will be more difficult.

Clean it Up …

Strip the body. The preferred method is a trip to your neighborhood sandblasting joint. Chemical dips work, but the solution tends to hang out in the little creases and crevasses within the frame only to slowly leak back out after new paint is applied, cracking it and causing it to peel off.

Bang it Out …

After your frame and panels are paint-free, the body work begins. Because the body shop industry is employed mostly by the insurance industry these days, a good body expert is hard to find, especially one who specializes in scooters and who is not addicted to Bondo. Body work is all about the little details. The ancient technique of using a hammer and a small, anvil-like "dolly" to pound out dents and dings and reshape the metal is becoming a lost art. Only a few scooter fanatics offer this service at an affordable rate. A restorer might also seek the assistance of a nearby auto or motorcycle specialist or learn how to do it themself.

Make it Fit …

After the body work is complete comes one of the most critical and overlooked steps in the restoration process: the dry assembly of the scooter. Before a frame and panels can be painted, make sure they all fit together properly. If they don't, the new paint will be far more likely to get scratched during final assembly. This is the time to fit up

The cases must be finished with the proper blast media and assembled using all of the proper gaskets, seals, and bearings. The slightest oversight and the engine will fail; if the engine fails, the restoration fails. Many options are available through catalogs and scooter shops that offer greater engine performance in the form of bolt-on horsepower, but a purist will tell you that an altered engine is not a restored engine.

trim pieces such as floor rails, badges, and legshield trim. Test the fitment of the cowls, the fork, and the fender, and deal with any unexpected problems before crunch time.

Paint it Up …

Once your body, fender, and panels are dent-free, crease-free, and perfectly sculpted and all of the finishing work is done, it is time for primer and paint. Painters are a strange lot. The painstaking precision required makes them that way. The chemicals don't help, either. Like body folk, most quality painters have grown fat off the insurance trade and have precious little time to undertake such endeavors as scooter restoration. That leaves the enthusiasts. A good paint job on a vintage scooter starts at around $800. Two-tones and other multi-color schemes can cost upwards of $2,000. The paint process is the most expensive aspect of a restoration, and the most critical.

Build the Motor …

While the frame is away for straightening and painting, the engine becomes the restorer's focus. It is one thing to rebuild an engine and another to restore it. All of the proper bits must be used, they must be plated correctly, polished, and they must be factory bits, not aftermarket.

Run the Wires …

With a freshly painted frame in hand and an engine that's polished, plated, and completely rebuilt, the re-assembly process may begin.Start with replacing of the scooter's nervous system: the wiring and the control cables. The cables and wiring harness must be routed and fitted properly. If they are not, frustrating repairs might need to be made after the final assembly.

Trim it …

Trimming the scooter is the final and most rewarding stage of the process. Make sure the engine performs well, and then install the side panels and the seat, followed by the rubber parts, the floor rails, and the badges. Patience is a must. The little details are the key to a successful restoration, and trimming is all about the little details.

Stick a Fork in it …

With the wiring and cables in place, the process of installing the front fork, fender, and headset may begin. New steering bearings are put in place and the whole configuration is merged with wires, cables, and switches to create the brain of the scooter before installing the engine, the heart of the scooter.

And the Engine …

Engine installation can be a tricky process when a freshly painted scooter is involved. One wrong move and that gorgeous lacquer is bound to get scratched. Great care must be taken to place the engine without banging it on the frame. The basic installation is very simple. The engine hangs from two points and, once the hardware is installed, the engine is in place. The difficulty comes in integrating the engine with the scooter's other critical systems: the nervous system (cables and wiring), the coronary system (fuel delivery), and the respiratory system (air intake).

Tune it …

Just when it starts to look like a scooter the reality sets in: the restoration process is nowhere close to over. This phase makes even veteran restorers pull their hair out in frustration. Making a scooter actually *work* sometimes proves to be, well, a problem. Elusive flaws vex. The carburetor must be properly tuned and adjusted, along with the ignition timing and control cables. Adjustments become unadjusted. It is the delicate process of putting the icing on a cake, having it collapse, making a new cake and having to ice it all over again. Love and care at this juncture make the restoration a success.

Finish it …

It's painted, it's put-together, and it runs. The scooter is almost ready to ride. This is the time to accessorize, the time for long-range test runs, the time for mounting the mirrors and those polished embellishments scrounged at the swap meet. It's where the rubber meets the road and the preparations are made for the unveiling.

Restoration: 1959 Lambretta TV175 Series 2
Owner: Dana Wilson

Few classic scooters pique the interest of enthusiasts like the venerable Lambretta Tourismo Veloce, or TV, models. Considered among the first performance scooters, the TV was a volley in the technological war that raged between Innocenti and Piaggio.

Lambretta enthusiast and collector Dana Wilson pined for one of the rare 175s for nearly sixteen years before a friend finally turned this one over to him. Having been neglected and left searing in the sun for decades, it was a project to be sure. Today, Dana's beautifully restored Series 2 machine is among a small handful of restored examples in the United States.

"The hardest part," Dana admits, "was the body work; pounding out the dents while preserving the beautiful factory stamped curves." He also preserved the original engine components. "It was a real thrill to crack that motor," he says. "As dirty as it looked on the outside, on the inside it was gorgeous."

Not everything on Dana's keeper is original. "I am a firm believer that there is no such thing as a factory restoration," Dana says. "I figure that as long as I am going to take one apart and put it back together, I might as well improve a couple of things."

Rite
UCATI

Chapter 3

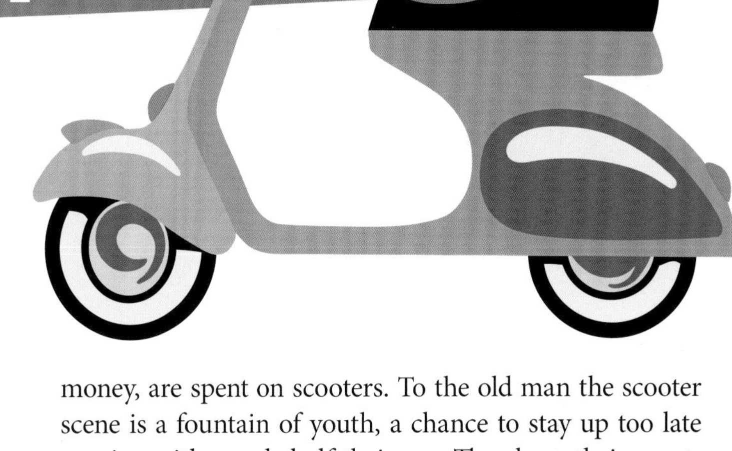

Can You See the Real Me

Scootering is an endeavor with extraordinarily broad appeal. There are all kinds of social reasons to ride them: rallies, coffee runs, Saturday rides with friends, flirting. There are also some practical reasons to ride them: scooters save gas, park easily, and reduce the wear and tear on expensive cars. Some people use scooters to pinch pennies, others pinch every penny so that they can spend them on scooters. The latter are the folks we're talking about here.

One of the many factors that make the North American scooter scene exciting is the vast array of colorful characters one encounters at rallies, races, and meets, not to mention local scooter shops. Scooterists are young and old, gay and straight, conservative and radical, male and female, skinny and fat, tattooed and not-so-tattooed, all sipping from the same cup—scootering.

What's Your Type?

For some, scooters are the basis of a lifestyle, while to others scooters simply fit neatly on the periphery. For those involved in a group identified largely by its fashion, such as mods, the scooter is sometimes an accessory, part of the look. For a scooter girl, a restorer, or a racer, the scooter is at the center of the culture. Other tastes, such as fashion and music, are secondary. Most of their free time, and extra money, are spent on scooters. To the old man the scooter scene is a fountain of youth, a chance to stay up too late rapping with people half their age. They keep their scooters clean and well maintained but they tend to focus their time on planning trips and organizing events rather than on acquiring more scooters.

Different elements within the scootering community value scooters somewhat differently, yet the little machines possess the power to draw people together. Scootering welcomes the masses—from yuppies to freaks, from rude boys to rockers, from ravers to hippies—in a wonderful way. It must be the two-stroke fumes.

It's almost impossible to draw a picture of a typical scooterist. Despite this, certain "scooterist stereotypes" have emerged. These were cleverly captured in a series of collector's cards stuffed into issues of *Scoot! Quarterly*. These "Scooter Stereotype Collector Cards" were illustrated by Dave Ross and captured for the very first time, a funny yet spot-on snapshot of some characters who help make up the North American scooter scene. Inspired by the cards, let us acquaint ourselves with some scooterists who loosely fit these stereotypes and some who resemble them to a tee.

The Mod

"We are the mods, we are the mods, we are, we are, we are the mods."

It is hard to describe as "modern" a social movement that has spanned four decades; yet, one cannot speak of mods without modernism. The term "mod" is an abbreviation of the word "modern," and the early British mods were a component of the modernist movement. Fashion was at the center of mod, and the clothes, indeed, made the mod. For men, their slick Italian suits, pointed shoes, and decked-out scoots were the latest and the hippest; a departure from the pastel preening of the 1950s and the "let it be" look of the middle 1960s. Women wore tight, knee-high, patterned dresses, with their hair short and boyish.

In the 1960s, mods emulated the adult middle class of the day in a sarcastic way. They poured attention into appearances and shunned the assumption that young people didn't need to work at looking good. They gained their societal independence by being noticable. They stood out first by looking sharp, then by latching onto scooters and sweet soul music.

Undoubtedly influenced by beat culture, British mods were awed by the swinging sound of American black music; first jazz, then the Motown soul. Sam Cooke, Aretha Franklin, Otis Redding, Jackie Wilson, James Brown, and their contemporaries provided the soundtrack of the movement. Mods danced the night away to these tunes while their American peers were embracing the Beach Boys and going psychedelic with British bands like The Beatles and The Rolling Stones.

The music evolved and a new mod sound developed in the U. K. Bands like The Kinks, The Jam, The Yardbirds, The Small Faces, and The Who, all of whom were heavily influenced by soul, helped the movement find its identity. A TV program called, *Ready, Steady, Go!* (which many consider to have been the best pop TV show ever) ran on the BBC throughout the middle sixties and brought "modernism" into British households, introducing teens to new artists and fresh sounds.

The ultimate "rocker," Gene Vincent dispelling stereotypes with a Vespa GC150

Scooters, particularly Lambretta TV and SX models and Vespa GSs, all very expensive scooters for the day, became the vehicles of choice for any mod worthy of a fuzzy parka. They needed scooters with lots of power. After all, they were toting a passenger's worth of weight in chrome and glass. Mods made their scooters stand out with varying arrangements of racks, crash bars, spot lamps, mirrors, emblems, and flags.

Now, in addition to fashion and style, the mods also had their very own nemesis. The natural enemy of the well-dressed, scooter-riding mod in the 1960s was the rocker. Rockers were influenced by Elvis Presley, Gene Vincent, and other Tennessee-style rock and rollers from the USA. They rode stripped-down Triumph and BSA motorcycles and, like the mods, tended to travel in groups or gangs. The two fashions were constantly at odds. Mods and rockers hated each other.

In 1964, tensions between the two groups reached a boiling point. On holiday weekend outings, the groups clashed violently on beaches along Britain's south coast including Brighton, Margate, Bournemouth, and Clacton. Many, many people were badly injured and many others were jailed. According to the BBC, literally thousands of kids were involved with more than 1,300 of them battling at Brighton. The atmosphere between the mods and the rockers was captured in The Who's famous rock opera, *Quadrophenia.* The album, and eventual film's intensity and wonderful music helped launched the mod movement internationally.

By the 1970s the mod movement thrived in the U. K. The ultimate mod band, The Jam, made the music that came from modernistic eight-track players and gave the mods a whole new identity. The Jam was comprised of three members, Bruce Foxton, Rick Buckler, and Paul Weller. They positioned themselves well outside the emerging punk scene and developed "mod" as an outside, yet closely related format. The band began

Love Rein O'er Me: Quadrophenia's Influence on Scoot Culture

Scooterists the world over cherish the album and subsequent film, *Quadrophenia,* by the British rock band The Who. The film's release in 1979 fueled the British scooter-boy craze of the 1980s and catapulted scootering as a counter culture across the Atlantic. It is still screened at scooter rallies from coast to coast; its soundtrack still pumps from sound systems.

It is the story of Jimmy and his mod mates in London who, maddened by mistreatment from rockers, fueled by drugs, youth, and music, and a disdain for society, help to turn bank holidays upside down during the famed bloody clashes of May 1964. While Jimmy is suffering the pains of adolescence—self doubt, alienation—his peers seem to be adjusting just fine to becoming adults. For Jimmy his acceptance as a mod, as a member of a tribe, is of the utmost importance—leaving his familial and personal relationships in tatters. When he discovers that his "tribe" may not be all he imagined, he is forced to confront and ultimately reject his adopted subculture.

Quadrophenia was more than a cult hit in North America, it was a catalyst, as it was around the world. Disenfranchised punks and ska music fans instantly identified with the characters and quickly the scooter scene began to emerge in cities and college towns. The mod craze spurned spin-offs—rude boys, scooter boys, skinheads. Young Americans began to see scooters as much more than economical little vehicles. Scooters became "scootering."

Tight Levis, Fred Perry sweaters, spots, horns, and, of course, mirrors help make the mod

releasing singles in 1976 with, "In the City," which Paul Weller wrote as a tribute to the punks. Sporting tight fitting Fred Perry golf shirts, tight Levis, and poochy parkas, the members of The Jam helped to popularize the mod look as we still know it today.

These days, mod is all about hanging onto a piece of the past while embracing the future. Most of today's mods have distinctly vintage tastes yet indulge heavily in technology—many are computer programmers and web designers. Still dedicated followers of fashion, mods are throwbacks who constantly work at keeping the scene fresh and jumpin.' Not quite modern but modernistic nonetheless. While the Brit pop craze of the late 1990s helped provide new music for them to enjoy (bands like Oasis and The Verve were big hits with many), mods are still tied to the past. Their favorite haunts still offer soul nights. Their scooters still have spots and mirrors. Paul Weller still tours and produces albums. The dancing has barely changed. The Italian suit, the Fred Perry, and the parka are mainstays. So are scooters. Few other counter cultures outside of punk can boast such a consistent existence over the past four decades.

Inspired equally by *Quadrophenia* and his own Irish heritage, Bill Haines of Fresno, California, put together a prototype mod bike in his Vespa P200, with a tricolor paint job featuring the orange, white, and green of the Irish flag. Why take his scooters to the mod extreme? "I'm just an attention whore," says Bill. "Whenever I ride this thing, everybody asks about it, everybody sees it. Even Ray Charles could have seen my scooter. It's a big rolling thing that screams, 'Look at me.'"

Scoot Racer

Soon after the birth of Vespa and Lambretta, scooter racing became an exciting—and affordable—form of motor sports action in Italy. Like motorcycle racing and bicycle racing, scooter racing became a popular Sunday pastime all over the boot in the 1950s.

The French cranked the sport up a notch in the mid 1950s, organizing the twenty-four-hour endurance race known as Bol d'Or. The annual Isle of Man Rally in the U. K. was also the site of a race that followed a course that circled the island. In the U. S., NASCAR waved the flag for its first sanctioned scooter race in 1959, New York City. But scooters proved almost too speedy for their own good. They so often beat motorcycles in American Motorcycle Association sanctioned events that the organization once banned the infernal little machines from competition.

Scooter racing lacked the drawing power of other

Modern and vintage models alike compete at the Scooter Nationals

motor sports, however, and organized races were few and far between. The modern scooter racing circuit can trace its roots back to Riverside, California, in 1986. Fabio Ballarin of San Diego's Vespa Supershop started putting together scooter races, piggybacking with a motorcycle racing organization to make it financially viable.

In 1990, West Coast Lambretta's Vince Mross built upon Ballarin's exploits when he founded the American Scooter Racing Association (ASRA), the granddaddy of scooter racing organizations in the U. S. Most of the younger scooter-racing bodies look to ASRA's rulebook as they write their own.

At first, about a half-dozen ASRA races took place in California annually, with five classes of machines: small frames, stocks, specials, automatics (under 80cc), and automatics (over 80cc). Held in conjunction with the California Motorcycle Road Racing Association (CMRRA), ASRA races have a minimum of three scooters per race, which is typically a distance of ten miles (six to ten laps, depending on the track). By the end of the 1990s, ASRA added a race in Las Vegas to the schedule, the organization's first foray outside of California.

After ASRA got rolling, it took a few years for the rest of the country to catch up. In 1999, Joe Kokesh launched Mid-American Scooter Sports—MASS for short—and began organizing a few races in Illinois and St. Louis, in association with a motorcycle racing organization, Championship Cup Series (CCS), much like ASRA had done on the West Coast. Over the next few years, the MASS schedule expanded to eight to ten events annually, with races in Ohio as well as Illinois and the Gateway City.

The Eastern Scooter Racing Association was born after Phil Waters, owner of Pride of Cleveland Scooters, and others noticed that a lot of racers from the East were making "pilgrimages of biblical proportions" to take part in MASS events. The organization struck a deal with CCS, and launched in 2001 with races in Charles Town, West Virginia. The calendar typically includes six events between spring and fall annually in Ohio, Pennsylvania, Canada, and elsewhere.

Scooter racers in a staggered starting position

Scott Smallwood of Supersonic Scooters

While it's come a long way since the founding of ASRA, scooter racing is still in the shadows of more visible motor sports. Dedicated racers have continually pushed the sport forward, but the popularity of scooter racing "comes and goes," says Vince Chu, retired racer and onetime ASRA president. "Sometimes it's a strong year and sometimes it's a light year."

The year 2004 saw the inaugural Scooter Racing Nationals, held in Erie, Colorado, during the week before the Mile High Mayhem Rally in Denver in late July. An east-meets-west event, the Nationals pit the best riders from ASRA, ESRA, and MASS against one another. The ASRA riders (and unaffiliated Coloradans) outdid their eastern counterparts in year one, with organizers aiming to make this grudge match an annual event.

Champion scooter racer Scott Smallwood fell in love with the sport at first sight and bagged more than sixty trophies from 1999 to 2004. "You're taking a scooter that was meant to negotiate the city streets of Milan and riding it well beyond what it was intended to do," says Scott, owner of Columbus, Ohio's Supersonic Scooters. "The opportunity to take a scooter and ride it as a road racing machine is quite spectacular."

Scooter Girls

Scooter girls are in the scene for the same reasons as the boys—they love their scoots, and they love being seen on their scoots. Some, particularly mods, preen and dress pretty; others go for the jeans and the flight jacket look, with maybe a tad bit of grease behind the ear. Still some can rarely be seen without full leathers and a massive Arai helmet clutched in one hand.

It's true that the desired "look" can be incredibly important to some women, going so far as to match scooters with shoes and handbags. But mostly it's about the

Missi Kroge, scooter girl extrodinaire

ride, the machine, and social scene to scootering. Some scooter girls do all or most of their own mechanical work, while others may have nothing to do with it, trusting local shops or buddies to handle the tweaks.

Scooter designers always considered women when sculpting the shapes of their creations. Scooters are designed to appeal to a woman's eye and sensibilities as well. The pre-eminent scooter girl scooter is the Vespa Primivera 125.

A prototype scooter girl, Miami's Missi Kroge went as far as getting married on an Italian vacation that included Eurolambretta, the premier international Lambretta rally. "I have always thought about how I look," Missi confesses. "I usually pick an outfit that's both cute and comfortable, you know, just in case I break down."

Chelsea Lahmers of SCOMO loves scooters as much as scooters love her

Though her feminine wiles have earned her legions of adoring male fans, Chelsea Lahmers of Richmond, Virginia, is on the other end of the scooter girl spectrum from Kroge. "I was never trained in the black art of girliness," she admits. Usually dressed in jeans and a t-shirt, with scooter grease beneath her finger nails, Chelsea is more at home with a wrench in her hand than she is with a tube of lipstick.

The Scooter Boy

The scooter boy is a broad category in this day and age. Basically, if a guy is into scooters pretty big-time and he is not a mod or a rude boy, then he's simply a scooter boy. To rein the term in a bit, however, and to get a little better handle on things, one can classify a scooter boy as "one who attends scooter rallies and accumulates event patches on a garment of some kind." While the flight jacket remains the garment of choice for patch collecting, many scooter boys have made the switch to mechanic's jackets, leather motorcycle jackets, even parkas. Scooter boys are involved in scootering as a hobby, a counter culture, and as an endeavor.

A scooter boy does not, as a rule, pay anyone to work on his scoot. The mark of a true scooter boy is a blistering burn earned in a late night roadside battle with a hot, broken exhaust pipe. He owns tools and he is not afraid to use them even if he and his scooter would be better off if

Grooming a new generation of scooter boy,
Ryan Basile and his son Gauge Innocenti Basile

he didn't. Regardless of his attire and his level of hygene, some grease will be visible somewhere on his person.

Scooter boys attend rallies outside of town and, while riding around town, make scootering visible in the community. Many go to bars while others spend most of their time in garages. These days a lot of scooter boys are daddies, too; busy bringing up another generation of scooter kids.

Case in point: Ryan Basile of Las Vegas, Nevada. “It's never been about fashion for me,” explains Ryan. “It's about scooters.” Ryan puts his money where his mouth is when he named his son: Gauge Innocenti Basile. Innocenti? Yes, the baby Basile's middle name is a nod to the manufacturer of Lambretta scooters. Will Gauge become a scooter boy, too? “Oh, hell yeah!”

And some scooter boys treat their scoots like children. The fanatical Darrin Lopez works on scooters for a living with his On the Road Scooters in Richmond, Virginia. “If you want me to fix your scooter you have to let me fix it front-to-back,” he says. “I probably net about $3 an hour for my work, because I cannot bill for anywhere near the time I actually spend getting things right.”

The Old Man

The old man scooterist hails from when scooters weren't cool. They bought them when the original dealerships were still around; they rode with the early clubs; and can tell some damn good stories from the early days.

Some of the old men of the scene are still very active in the rally circuit, travelling from place to place to show off their rigs and chew the fat with the young people. One of the most popular annual events among the old men is the Amerivespa Rally, held each year in a different city by the Vespa Club of America. The balance of younger and older scooterists at Amerivepsa is one of the rally's draws. The old man tends to be more interested in the machines themselves than in some rally activities, and at Amerivespa the Vespa is front and center. It gives him the opportunity to impart his knowledge on the boys and girls of today.

The old man's energy was critical in the early stages of many scooter scenes, even if he is less involved now. Scooterists in every scene cherish these elder statesmen and have much to thank them for.

Two well-known scooter stalwarts who fit the old man description to a tee are Randolph Garner of Cleburne, Texas, and Waid “Scooter Daddy” Parker of San Diego, California. Garner started the Cushman Club of Texas in 1983, then served as the president of the Cushman Club of America from 1987 to 1988. In 1995,

This clever license plate belongs to Walter Garson

he founded the Vespa Club of America. "I was going to call it the European Scooter Club of America, but that sounded a bit too weird," he remembers.

Waid earned his "Scooter Daddy" nickname in the early 1990s when he decided that he would do everything he could to make sure that scooterists in San Diego never had to leave their scooters stranded. He printed "Scooter Daddy" business cards and instructed enthusiast to call him anytime, day or night. Today, Waid's Ford rescue van has more than 300,000 miles on its odometer.

New School

Not every scooterist prefers a vintage machine. In fact, the number of new school scooterists in the North American scene will someday eclipse that of vintage riders. Modern scooters already out-numer old ones and many owners of these twist-and-go models are becoming active in the rally circuit. Models like the Vespa GT200 and the Kymco People 150 offer enough power to keep up with, even zip past, the fastest vintage bikes, and the styling keeps getting better and better. Manufacturers like Kymco and Genuine Scooters donate machines to the raffles at scooter rallies in the hopes that vintage enthusiasts will soon embrace their machines.

New schoolers are less likely to do mechanical work at home. Most of them have warranties, after all. The most enthused among them, however, turn to the aftermarket for tuning parts, bolt-on accessories, and chisel on their scoots every bit as much as the old schoolers do.

Many design custom vinyl graphics sets and personalize their scooters in many ways. They ride with pride and with all the enthusiasm of a true scooterist. They also travel in increasing numbers to rallies around the country. Vintage freaks are even becoming new schoolers as, in the quest for fun and the search for performance, they have discovered the wonders of scooter technology.

New schooler Bryce Ludwig of Lawrence, Kansas, is ultra-enthusiastic about modern scooters. His own ride is a 2003 Peugeot Looxor, an unusual 150cc tall-wheel model that was not widely distributed in the U. S. His interest in modern scooters goes far beyond his own mount, however. Bryce test rides almost every model he can get his hands on and does new model reviews for *Scooter World* magazine. He finds that scooters posses traits that, as a designer, he covets.

Beyond the Cards

Beyond the universe of Ross' trading cards, the stereotypes for scooterists mount. Here we have some additional scooterist types and the folks who emulate them.

The Restorer

There are special people in this world that can look at something hideous and see beauty. They know what it means for something to be fundamentally sound yet optically unappealing. They can pick something apart in their minds, piece by piece, and see what it needs, what it is going to take to make it what it once was. They are the restorers.

A restorer is, by nature, a highly organized individual. They can also assemble a complicated array of parts and fasteners in a cautious and careful way. They measure twice and cut once, with concerns lying deep beneath the surface of their work. They know the flaws that you cannot see and they refuse to accept them. Even the primer beneath the paint must be the right color. Each nut and bolt must be of the correct factory finish and every color must be true.

A restorer collects knowledge and information in order to assure precise assembly. They accept no fillers, no shortcuts, no mistakes. Loyalty is to the factory, to the one who designed the machine. The restorer does not consider a scooter restored if it is painted pretty metalic colors and is adorned with chrome and jewels. It a travesty to intentionally alter, much less drill into or cut down, a vintage scooter. The factory did it the right way and nobody else.

This breed of scooterist patiently polishes, painstakingly plans, and produces only world-class scooters. North America is a hot bed for scooter restoration due to the large number of low mileage, rust-free scooters that are here and thanks to the wealth of experience throughout.

Tim Stafford, who opened San Diego's TJ Scoots in 1989 with partner Jay Tellier, is a restorer's restorer. He's done countless gorgeous restoration jobs, including numerous Vespas as well as a myriad of other vintage specimens. "To me, restoration is making a bike as technically correct as possible," Tim says. "In other words, it's trying to make the bike as close to how it came off the factory lines as possible, that's down to the finish on the hardware, the color of the primer, every little nuance."

The Hot Rod Builder

These enthusiasts have amassed a huge collection of tools. Scooter bits litter their garages, their basements, even their kitchens. Their annual carb cleaner budgets exceed their food budgets. They tune, tweak, fidget, alter, test, and retest. Hot rod builders believe passionately in the pagan gods of displacement and compression, and their brain waves can only be measured in RPMs.

Only a hot rod builder can explain the desire to turn a scooter, a machine intended to help save on gasoline while putting out at road speeds, into a fire-breathing, hot-rod beast with a pipe so loud it can shatter glass. It seems a bit counter-intuitive, but like VW freaks and people who race lawnmowers, they do it.

Hot rod scooter builders measure their success not in the beauty of their scooters but in numbers: numbers from the dynonomter, numbers from the traps at the drag strip. And nothing motivates the hot rod builder more than the idea of smoking a motorcycle in a street race or flying past the fatest scooters on the rally at seventy-five miles per hour.

The hot rod builder looks at their little machine with its ten-inch wheels and sees their own, inexpensive little racer. They pour over scooter magazines and scour the Internet in the race to stay on top of the latest R & D on scooter pipes and scooter pistons and scooter tires. They also think of ways to use motorcycle parts, even motorcycle engines, to make a scooter faster. The hot rodder could give a damn about the guy at the factory who designed the scooter. The guy was dead wrong; the antithesis of the restorer.

Casey Earls climbed out of her tent on her twenty-ninth birthday at the Tucson-Nogales Rally in 1997 with a hangover and the determination to do something with her life before the big 3-0. And that she did. Within a mere sixty days, she collaborated with her boyfriend Barry Synoground and started fleshing out the concept for a quarterly magazine dedicated to vintage scooters and the culture that flourishes around them. An "ass-ton" of hard work later, *Scoot! Quarterly*'s first issue ran off the press in August 1997.

Soon after the launch, diehard scooterists all over the U. S. accepted and adopted *Scoot! Quarterly*. The magazine became central to the burgeoning national scooter scene, and gave scooterists a way to communicate with each other, scooter shops and rally organizers a means of advertising to the national market, and the entire community a venue to celebrate the shiny little machine that was at the center of it all. Mechanical eye candy defines the magazine—gorgeous centerfolds that make hardcore scooterists drool.

After publishing twenty-six issues, Casey and Barry decided to sell the magazine—the magazine's demands were cutting into their sleep schedule and just about everything else. "It was too much work," says Casey, whose day job was running a San Francisco nightclub with Barry. Thus enter the trio of Bay Area scooterists of April Whitney, Josh Rogers, and Mike Zorn, who came together and bought the magazine.

April had written for the magazine previously, and all three knew Barry and Casey through the local scooter scene. April says that it took a few issues for the new crew to get their feet wet. "It's been educational," she says. "I didn't have a full idea of what went into making a magazine."

"*Scoot! Quarterly* remains organic and not corporate," April adds. "The people who read it contribute to it. *Vogue* needs the readers to buy the magazine; we need the readers to make the magazine." The main change the new publishers want to affect is a bit more focus on new scooters, because the market is booming and also because "there's a finite number of vintage scooters," April explains. "However, all three of us ride vintage scooters. We're trying to balance it out."

And it's still a labor of love. "We're all kind of scraping by," says April. "But subscriptions are up and 10,000 copies continue to reliably roll off the presses four times a year—on deadline."

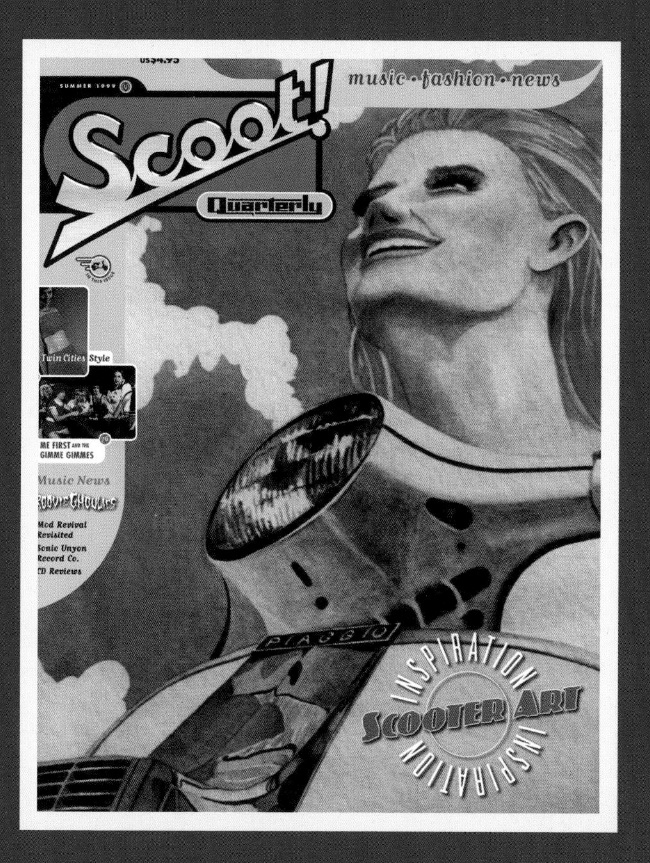

Denver's Bottle Rocket Scooter Club

Join the Club:
A Small Sampling of Scooter Clubs

While scootering is a hobby steeped in individualism, it is all about friendship. Joining a scooter club is a big part of it for many, many scooterists. Clubs provide both technical and personal support. Club-mates spend evenings together at a pub or on the cold, concrete floor of a garage, building projects and making repairs both large and small. They communicate with one another by phone, in person, and on the web. They rally together.

It would be nearly impossible to list all of the scooter clubs in North America. Some have as few as two members! But the following group of scooter clubs represents some of the most well recognized and well organized on the continent.

National Scooter Clubs

Jedi Knights

Easily the largest scooter club in the USA (aside from the Vespa Club of America), the Jedi Knights were "fired up" in Ann Arbor, Michigan, in the mid 1990s by a small group of friends, a few of whom attended the University of Michigan. The club's early slogan was simply, "World Domination." Boasting "millions and millions of members," the Jedis have absolutely swept the nation with large, active chapters having formed in many major cities including New Orleans (Dagobah Chapter), New York, San Francisco (Cloud City Chapter), and Denver (Naboo Chapter).

"Everywhere we went," says Scott Hyland, one of the original Ann Arbor members, "people identified with what we were doing, which was having fun. Instead of taking ourselves seriously we were embracing the nerdiness of the whole thing. I mean, here we are, a bunch of dorks on 'skirt bikes' celebrating *Star Wars* and waving plastic light sabers. People from all over the country

Jedi's mark their territory

wanted to get involved. They said 'Hey, we're dorks, too, and we like *Star Wars*.' It's the hilarity of the whole thing that makes it fun."

Jedis must be "knighted," a ceremony involving the existing Jedis christening the member-to-be with their light sabers. To be considered for membership, one must simply express the desire to become a Jedi. Knightings must be approved unanimously by local Jedis and are performed only at scooter rallies.

"Having already grown by leaps and bounds, the club is changing," adds Scott. "We are moving away from world domination and more toward the general firing up of things." The Jedi Knights have won more "Best Club Turnout" awards at more rallies than any other club in the history of the American scene and have also made an indelible mark on scooter racing. They probably use *The Force*.

Pharaohs

San Diego was a hotbed for scootering in 1993 when a U. S. Marine named Norm and a handful of friends founded the now national Pharaohs Scooter Club in Oceanside, California. "They were just three guys who wanted to get together and ride," explains "Grandpa" Nate Frazier of the San Diego chapter. "The club has never been about drama. Our slogan is basically, 'Just get on your motherfucking scooter and ride.'" Nate adds that he, "would never want to be a part of a club that only gets together once a year for the rally." For him and the other members of the Pharaohs, the club is like an extended family. "If a club-mate calls me at 2 am because he's broken down, I say, 'Alright,' and I jump in my truck. It's that simple."

Chapters have since cropped up in other parts of the U. S., including San Francisco, Chicago, Salt Lake City, El Paso, Madison, and Phoenix. "I kind of started up the Phoenix chapter when I moved out there in 1995," says Nate, "but I hate to toot my own horn."

The club has become known for its rally antics. Pharaohs are legendary for burnouts, wheelies, and flaming ramp jumps. Some of the members are extremely talented riders with the ability to wow rally-goers. "Our reputation is that we are all a bunch of hooligans," says Nate. The club is dedicated exclusively to vintage scooters and all of the focus of the group is on riding the machines and traveling together

Southbay Scooter Club

to rallies. The club is responsible for two major events, the King Tut Putt, which occurs in San Diego in March, and Too Fast For Love in Phoenix in May.

Secret Servix

Established in Denver, Colorado, in the late 1990s, the Servix is the largest all-girls scooter club in North America, with members on both coasts and many places in between. The club's name is a parody of California's most famous club, the Secret Society. And like the name implies, Secret Servix is also a celebration of femininity.

The sexy scooter club gained fame early in its existence. The girls' sharp outfits, cool scooters, and warm demeanors made them the center of attention at nearly every rally they attended and they proved adept at making friends. Members are active in planning the Mile High Mayhem Rally in Denver each year and assume the task of leading the annual Friday shopping ride, one of the rally's most popular activities.

Each girl is equipped with a purse full of silk-screened ribbons that they may present at their discretion to scooterists who exhibit style in virtually any manner. The ribbons are quickly marked with a Sharpie when style is spotted. Past awards have included "Hottest Pregnant Chick," "Cutest Shoes," and "Least Likely to Get Lucky with a Servix Girl." The Denver rally serves as an annual meet-up for the club; many members have relocated from the city so Mayhem is a homecoming. According to club co-founder Missi Walker, the Secret Servix has only one rule: "We don't steal each others' boyfriends."

West Coast Clubs

Rally Kings, San Francisco, California

One of the country's best known scooter clubs, the Rally Kings are responsible for some of northern California's biggest and best rallies, most notably Kings Classic, a massive annual event that started as Scooter Thing in the early 1990s.

These days, the club's membership has dwindled significantly and it exists largely to organize the Classic. "That's pretty much all we do anymore," says club member and San Francisco Scooter Centre owner, Barry Gwin. But the legacy of the good old days lives on through one of America's largest scooter rallies.

Juggy of the Burgundy Topz

Secret Society, San Diego, California/San Francisco, California/Boston, Massachusetts

Secret Society is one of the most famous clubs in North America and is considered to be the oldest existing group in the United States. Founded in San Diego in 1983, the club currently has additional chapters in San Francisco and in Boston, and is responsible for a number of annual events but largely for Scooter Rage, their monstrous national rally that was once held in January but has been shifted to June of each year.

Rage, along with Kings Classic and High Rollers Scooter Weekend, is one of the largest events in the country, often attracting well over 300 scooters. The club is part of the Northern California Scooter Council and its members are very active rally-goers.

"One of the requirements to be considered for membership is to own a metal scooter and that's where the similarity stops," says San Francisco member Mike Zorn. "We are about as diverse as the bikes we own and ride."

Burgundy Topz, Sacramento, California

In the mid 1980s, as the scooter scene began to gel in northern California, a group of enthusiastic scooterists started getting together in the back room of an ice cream joint in Sacramento. The group gradually became a scooter club but only in a very informal way. Members credit this casual origin with the longevity of the Topz. Consider this excerpt from the club's web site: "No President, No Treasurer, No Sergeant at Arms, hell, NO Dictator. This is how we, Burgundy Topz, have survived for almost two decades."

They tried in the beginning to form some structure but members realized that such efforts cut into their fun. According to a club profile in *Scoot! Quarterly* in April, 1999, "Our structure evolved based on regular members and we cemented a lifelong brotherhood composed of people more interested in hanging out and riding than just pissing away time trying to elect presidents and kings and sergeants and whatever else."

Burgundy Topz Scooter Club, 1986

Burgundy Topz have been responsible for two major events for about the past decade. They are Scootouring, a camping event held every spring in the foothills of northern California, and Worshiping the Beast, an urban autumn event that features a custom scooter show as well as beautiful rides through Sacramento. The club is involved with the Northern California Scooter Council and its members attend rallies up and down the West Coast. Potential members must attend many meetings, usually over the course of a year, before they will be considered for membership. Once considered, they must be approved by unanimous vote. As a result the club remains smallish, with around twenty members, but the group stays tightly knit.

Wussys, Pacific Northwest

It is not all that uncommon for members of scooter clubs to get the same tattoos emblazoned on their epidermis. To become a Wussy, one must get a gigantic one that says "WUSSY" in the classic Old English font. Most members have the tattoo across their chests in massive letters. Why? "After much tequila, Zima, and lost rounds of video poker, there was a scene much like something out of *Beavis and Butthead.* These guys had a discussion about scooters, chicks, and how they were a bad influence on each other in general. The conversation drifted to the fact that they were tired of people (meatheads in general) trying to pick fights with them any time they went to a bar and just wanted to hang out."

The explanation continues: "America's concealed weapon laws were discussed, and the consensus was that bars, weapons, and meatheads don't mix *at all.* With no desire to be a part of the aforementioned concepts in any combination, the idea of really big Old English tattoos fell into place. How hilarious would it be to have a *really* big tattoo—the pinnacle of toughness—that said something that completely defeated the purpose?"

They decided upon the word "wussy." Hilarity ensued. "Our boys couldn't stop laughing at the concept of the biggest, toughest tattoo that completely defeats the very purpose of what the average tool thinks getting a tattoo should be about."

The group began in Portland and quickly spread to Seattle as well as parts of Canada. Members address each other as a number, referred to as one, two, and so on, in the order in which they joined the club. The Wussys are not strictly a scooter club. It just so happens that most of them ride scooters. They state that their members have, "carted their machines around North America to wreak havoc on the hearts and minds of decent people everywhere."

Hard Pack, Orange County, California

Very few scooter clubs in the U. S. can claim to have as many downright bad-to-the-bone scooters among its ranks as the Hard Pack Scooter Club. This group, which includes legendary scooter builder Nyle Schafhauser, is proud that its members are seen on some of the most awesome Vespas and Lambrettas in California. According to the club: "A large majority of Hard Pack members ride scooters that have tuned high-performance motors. Much of the technical scooter wizardry comes from behind the scenes. Three-time ASRA champion Nyle Schafhauser provides members the necessary experience and knowledge fundamental in building a fast, consistent, and properly running scooter."

The Hard Pack began in 1992 and has hosted a great many rallies in southern California. In the early days

the club hosted an event called Club Day that evolved into their big annual event, Orange Crush. The club also does its part for charitable causes, hosting a toy run each December. The Hard Pack consists of around twenty-five members, some of whom no longer reside in California.

Twist & Play, Portland, Oregon

Portland is no ordinary scooter city. It is home to one of the most eclectic and bizarre scenes anywhere. Portland, like its neighbor Seattle, is a liberal city where almost anything goes. Twist & Play Scooter Club, founded in the mid 1990s, is scooting proof of that. This zany cast of characters would never want to be caught taking anything too seriously.

Twist & Play hosts such local scooter events as the Dirty Clown Run, where riders terrorize the city dressed as insane clowns, and the Poke-N-Dragger, involving male scooterists riding in drag to various bars and creating a scene at every one. According to member Dave McCabe, "We kind of created a tone of being jackasses."

The Oregon Scooter Club and Vespa Club of Oregon is open to any scooter model and to people of all ages and backgrounds

The Los Gatos Gordos Scooter Club

Ratty scooters are one of the trademarks of the club. "Our bikes don't have to be pretty," Dave explains, "but they have got to run. Actually, if they are really beat up, that is kind of a bonus."

Twist & Play plays host to its Pacific Northwest neighbors each spring with its big event, Spring Scoot, which draws anywhere from fifty to seventy-five scooters. Most club members also make it a point to visit annual rallies in Seattle, Vancouver, Victoria, and Eugene. "The entire Northwest scene is super tight," Dave says. "We have our own legacy of rallies out here so most of us don't really go to California too much."

Hell's Belles, Portland, Oregon

Portland's first all-girl scooter club demonstrates that necessity is the mother of invention. The club was formed in 1995 when scooting girls in the city began to find that the local scooter clubs did not accept them. So three women got together and started the Belles and

Members of Los Gatos Gordos on a ride

by 1996 the club was hosting its own rally. Men have always been welcome to attend the Rally from Hell.

"Considering that usually this rally has been organized, promoted, and carried out by a small group of gals (usually less than five), it's a pretty cool achievement that we've lasted so long." The women also organized an event called Porn and Candy in 1999, which was hosted in a house often used by a swingers club.

Los Gatos Gordos, Pacific Northwest

The Portland and Seattle scooter scenes are very closely intertwined, yet few clubs have chapters in both cities. Los Gatos Gordos, the only scooter club in the country that caters to Spanish-speaking members, is the exception to this rule. The club was started in Seattle by two Latino scooterists who noticed that being Hispanic in the mostly white scooter scene caused them to stand out a bit, so they decided to play it up. The club slogan reads, "If you fight one bean, you have to fight the whole burrito." And the club's members each have a nickname, most of which—like "El Dorko," "Pablo," "Diablo," and "Lil' Guapo"—are distinctly Spanish, albeit tongues firmly planted in cheek.

Middle America Clubs

Upstart, Salt Lake City, Utah

Salt Lake City might be better known for the Mormon Tabernacle Choir than its scooter clubs, but the underground of Brigham Young's fair city is home to one of the most bustling scooter scenes in the U. S. Bolstered by impressive scooter sales figures, Salt Lake is quickly becoming a mini scooter mecca.

The SLC scooter scene happily played host to Amerivespa 2004 and the Vespa Club of America's national rally. The Upstart Scooter Club, a loose collective of area scooter hounds that began in 1999, also throws a large-scale event each September. Upstart lends its assistance each spring to the independent scooterists who organize an annual rally in Moab, Utah. "We try and help those guys out as much as we can," says Larry Taylor, one of the oldest Upstarts. "Moab is my favorite rally," he adds.

Brigham's Bees, Provo/Orem, Utah

Founded in 2001 by three scooterists, Brigham's Bees Scooter Club quickly expanded to more than thirty members. "Being in a scooter club has kind of become the hot thing to do in this area," says David Hurtado, club co-founder and owner of Provo's Scooter Lounge. "Being involved in scootering has become a real part of the scooter-buying experience for people. It's great for business because people aren't just buying a product, they are investing in a lifestyle."

Indeed, the Provo area has become a genuine hotbed of scooter activity. "Our market is particularly rich for its size," says David. "Stella sales have been really strong here, which is cool because that's the perfect bike for runs and rallies."

While Brigham's Bees are not in any way associated with Brigham Young University, several of the members attend the school. David believes that the school has been a boon to the area's scootering community. "The scooter scene is actually getting younger here thanks to all the students," he says. "It's cool because it seems like most scooter scenes are aging. I think its bodes well for the future here." The Bees host their own annual rally, called Provophenia, which draws riders from all over the state.

Thomas Heath of ACE Scooter Club and a few of his scooters

Bottle Rockets, Denver, Colorado

"Bottle Rocket is Go!" was the slogan the club chanted when it was born in the winter of 1999. Started by a group of young professionals, most relatively new to the hobby, the club has moved from the onset to build nice scooters and to enjoy scootering in a stylish and responsible way. The club's stated purpose: "Dedicated to the riding, maintenance, and restoration of classic scooters. Devoted to the aesthetics of the scooter, and the camaraderie of those who ride them. Committed to the social responsibilities, public image, and personal safety of motorscooter riders."

With roughly twenty members, Bottle Rockets is one of Denver's most traveled and active clubs, and many of its members ride spectacularly restored bikes. The Bottle Rockets contribute to the organization of the Mile High Mayhem Rally and are, appropriately, mostly in charge of rider safety.

Members of the Bottle Rockets Scooter Club at the Gothic Theatre in Denver

Mods and Knockers, Denver, Colorado

"We love drinkin', smokin', wearin' short skirts, talkin' shit, and cuttin' bitches while riding our hot scooters!" explained a spokesperson for Denver's wildest all-girl scooter club to *Westword* in a July 2004 interview. Established in 2003, the group includes members with monikers like Super Suzi, Galaxy Girl, Kittenpanties, Dr. Bonermaker, and the Tarynnator, and they are all hardened, dedicated scooterists who love to lose control.

Mods and Knockers girls pile up the miles, riding their scooters almost every day. Community service is also on the Mods and Knockers' agenda. The club has hosted a fundraiser for a rape crisis center and organized a voting drive. They also travel very well and represent the Denver scene at rallies throughout the region.

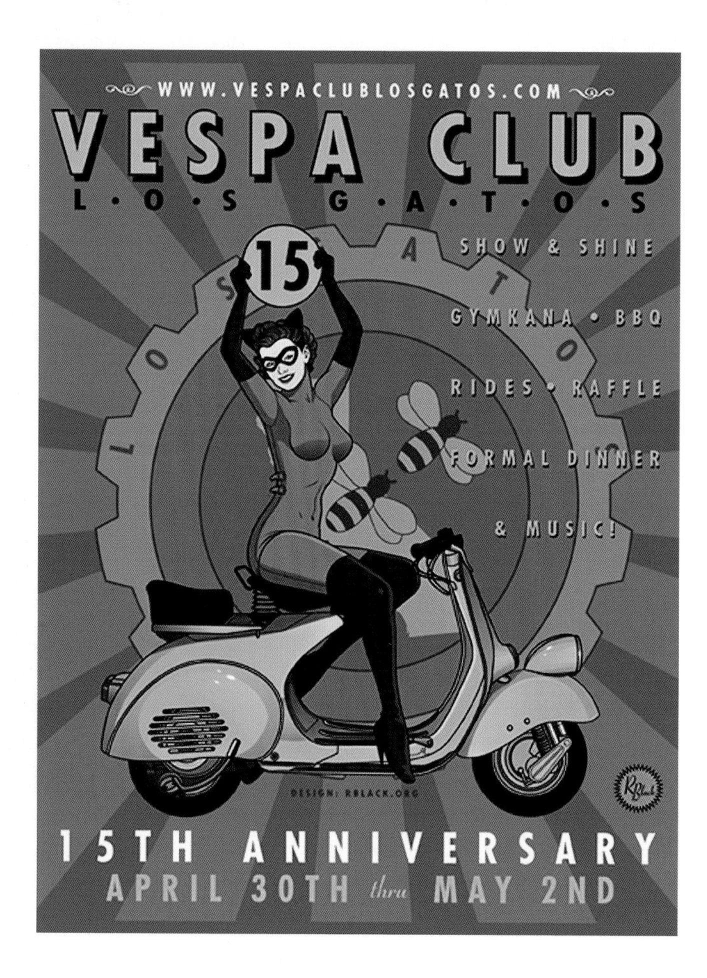

Blue Smoke, Denver, Colorado

The slogan of Denver's original scooter club, Blue Smoke: "All we want to do is ride our machines and not get hassled by the man." The club, which insists that, "your future is our past," consists of five members. None are ever added and none are subtracted; membership is for life. It was started in a fraternity house near the Colorado State University campus in the early 1990s. "There were always scooters parked out front of the house," says founding member Phil Lombardo, the "Mayor" of DCD (Denver City, Denver), and chief organizer of the Mile High Mayhem rallies. "We held our first rally in [Denver's] City Park."

That rally, held in 1993, and the other rides Blue Smoke organized around that time acted as catalysts, igniting interest in scootering in the Denver area. "The Pub Scouts were getting together at around the same time," says Phil, "but we were completely unaware of each other until one day when I was handing out fliers for a ride at Paris on the Platte [a popular Denver coffee shop] in '93 and we ran into Adam Baker from the Pub Scouts." The Denver scene was, in essence, born on that afternoon. Soon, ACE Scooter Club also formed, and by the summer of 1994 the scene was among the most active in the country.

ACE, Denver, Colorado

With an official club slogan like "Cocaine, scooters, and whores," ACE is the real deal. One of the Rocky Mountain region's first and most important scooter clubs, ACE was founded in 1990 and has varied in size over the years. The club's current incarnation is its most ruthless: the dozen or so active members are uniformly hedonistic and unruly. The group provides entertainment each year at the Mile High Mayhem Rally by booking musical acts, spinning records, and unveiling clever T-shirts with slogans casting dispersions on other people and clubs. These phrases have included: "Big Dave Sucks," "Lots of Bottles, No Rockets," and "Shifty, but Never Shiftless," the latter two being slams on other Denver clubs.

Fort Collins Scooter Gang, Fort Collins, Colorado

The I-25 corridor is lined with scoter clubs from Albuquerque to Fort Collins, where you can find an active group of scooterists known as the Fort Collins Scooter Gang. The Gang's slogan, "Clubs are for Mouseketeers," is written in red lettering across the bottom of its logo, as is the altitude of the club's hometown: 4,984 feet above sea level.

"There are a lot of new kids coming onto the scene," says Nick Andermann of Fort Collins' Unauthorized Vespa Shop. "A lot of them just look at scooters as basic transportation." There is an interest in vintage scooters, though, and nostalgia is a big reason. "A lot of the older folks up here like the vintage scooters because they had them when they were kids. They don't usually join the gang, though. It's not really their thing."

Atomic, Albuquerque, New Mexico

Amidst the lovely scenery of southern New Mexico sit testing ranges onto which many bombs, some atomic, have been dropped in the quest to defend the free world. It is for this reason that Albuquerque's largest scooter club decided to call itself "Atomic."

Molded with a mix of scooterists old and new and supported by a strong local scooter shop, Atomic has opened new eyes to scootering while providing new spark to Albuquerque's established scooter scene, which began to wane a bit after 2001. Despite having several key scenesters relocate so such places as Colorado and California, the city maintained a solid core of fun-loving scooterists.

In past years, Albuquerque clubs have hosted a fantastic camping rally in the Jemez Mountains complete with hair-raising rides and memorable campfire gatherings. They have also shown *Quadrophenia* on the big screen at the city's historic El Rey Theatre. Atomic's goal is to pick up the pieces of these past events and make their town a draw for scooterists once again. One such event is Camp Scoot, a camping rally in the tradition of the well-liked "Dead Man" runs that were hosted by the now-disbanded Nomads Scooter Club. With about twenty-five members, Atomic has a bright future.

Peak, Colorado Springs, Colorado

After hosting Amerivespa two years in a row in the mid 1990s, a great deal of enthusiasm spread through Colo-

Founded by Salt Lake's Bill Swinyard, Scoot.net is an online library filled with information, links, and the very popular auto-submit photo galleries filled with rally pictures

Of course, the best part of being a member of a scooter club is always having someone to ride with

rado Springs for scootering and playing hosts to events. That, along with the growth in the number of scooter riders in the area, led to the formation of Peak in 1998. With its sunny summers and its majestic scenery, Colorado Springs is a great city for scootering, and each year Peak treats regional scooterists to Movin' On Up, its June rally that serves as a nice warm-up for Mile High Mayhem. Per the club's slogan, members "ride, restore, and maintain classic motorscooters."

Pride of Cleveland, Cleveland, Ohio

Once known as North Coast Scooter Club, Pride of Cleveland has been active since 1994. This wily group of scooterists does not believe in having meetings and is as tightly knit as it is loosely organized. "Unlike most scooter clubs, we have an informal scene. We keep the politics to a minimum with each member doing what they can to make the club better."

Perhaps it is for this reason that the club is cast so much in the image of its founding member. Phil Waters, Mr. Pride of Cleveland himself, is a fixture at scooter rallies from coast to coast and the charismatic proprietor of a scooter shop called, you guessed it, Pride of Cleveland Scooters. Part-time racer, part-time mechanic, part-time sales guy, and full-time great guy, Phil draws people to him. The result of this magnetism is that the Cleveland scooter scene has become vigorous.

Pride of Cleveland Scooter Club hosts a minimum of one major rally per year. In 2002, the club scored an all-access pass for their rally guests to the Rock and Roll Hall of Fame. Cleveland rocks and so do these scooterists!

Ten Year Lates, Cincinnati, Ohio

"When the end of the world comes, I want to be in Cincinnati. Everything that happens comes there ten years later than anywhere else." —attributed to Mark Twain

Needless to say, the Ten Year Lates Scooter Club of Cincinnati, Ohio, gleaned its unusual name from the aforementioned quote. The club, formed in 1999, is among the largest in the country in terms of membership. With nearly fifty active participants, the Ten Year Lates (a. k. a. the XYLs), pride themselves on accepting members with all types of scooters, old and new. "We would probably be two or three clubs if there were politics involved," says XYL Casey Beagle.

Instead, the unified collective of scooter riders from Cincy and neighboring Kentucky flock together to meetings on Thursday nights at the Comet Bar in the Northside neighborhood. "The coolest thing is to see the mix of people we have," says Casey. "At our meetings you are bound to see a middle-aged professional guy, possibly a doctor or a computer programmer, talking to a gutter punk kid about scooters. It's the scooter that bring us all together."

Once a year, the XYLs bring many, many people together for the region's largest scooter rally, WKRP in Cincinnati. Past events have drawn close to 150 scooters. "The rally is growing like mad," Casey explains. "It's tough to keep control of it."

Like many clubs, the XYLs benefit from having a member who owns a local scooter shop. "Metro Scoot is great," Casey says, "and a lot of us like to deal with Supersonic in Columbus for performance stuff."

Sputnik, Oklahoma City, Oklahoma

The explosion of scootering found in Oklahoma could be considered a bit of an anomaly considering the rural image, but the dedication of the Sooner State's scooterists is unbelievable. Sputnik Scooter Club held its first rally, Test Flight, in 2004, but a handful of the club's members have been very active in the hobby for a very long time. Scooterists from Oklahoma have always traveled well, making the scene in Denver for Mile High Mayhem, at Amerivespa events from coast to coast, and at rallies across Texas. A couple of them are also regulars at European rallies like Eurovespa, which is a rally like its American counterpart.

Continental Kings, Tulsa, Oklahoma

Oklahoma might not be the first place you think of when you think of scooters, but the state is home to a collection of very faithful scooter hounds. Among them is Tulsa's Zach Matthews, known to most scooterists as "Mobboss." The boss and his close cohorts make up the Continental Kings Scooter Club, which has about a dozen active members. Together with their friends in the Sputnik Scooter Club out of Oklahoma City, the Kings help insure that the state makes a fair showing at rides and rallies near and far.

Zach, who discovered scootering in the mid 1990s, has attended events in Italy, Spain, and Argentina, as well as some in Asia. He and his mates also make a strong showing at Amerivespa events wherever they are held. Zach says that Oklahoma scooterists consider themselves to be part of a larger scene that includes Texas and Louisiana. "It's kind of like the Denver/Salt Lake/Albuquerque connection," he explains. Every scooter scene needs strong leadership and Zach "Mobboss" Matthews is proud to provide it with the help of his club.

Back East

The Oppressors, Philadelphia, Pennsylvania

The black sheep of the Philly scene, the Oppressors Scooter Corps is a smallish club with around ten members. They share an interest in scooters as well as a taste for Pabst Blue Ribbon, the official beer of scootering. This quote from the club's web site explains the motives behind starting the club: "The Oppressors were really formed from the misfits of their local scene. So suffering from a severe persecution complex, and a wicked hangover, they decided to strike out on their own path." Apparently, following the herd is not this group's strong suit.

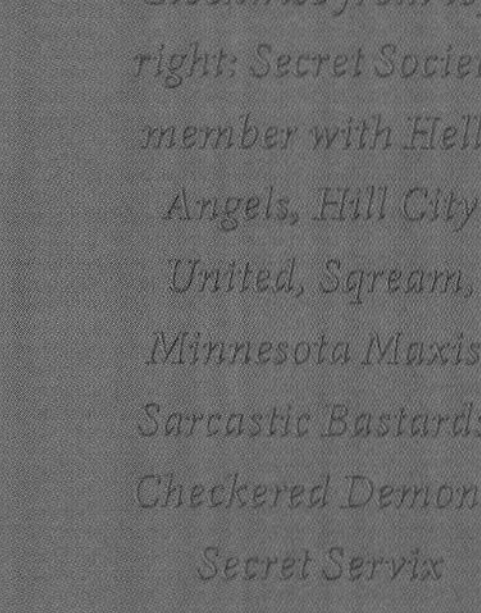

A few club photos. Clockwise from top right: Secret Society member with Hells Angels, Hill City United, Sqream, Minnesota Maxis, Sarcastic Bastards, Checkered Demons, Secret Servix

Lambretta

PITTSBURGH VINTAGE, PITTSBURGH, PENNSYLVANIA

Every scooter scene has a character without whom there would be nothing; someone whose energy and passion sparks interest from others on scootering, rallying, and generally everything else. In Pittsburgh, that person is Paul Hellfrich, co-founder of the Pittsburgh Vintage Scooter Club.

In the late 1990s, Paul convinced fellow Pitt scooterists to travel with him to such northeastern rallies as the popular Niagara event. In doing so he "set the hook." Interest not just in scooters, but in the scootering life, grew in the Steel City, and Pittsburgh Vintage was born. The club boasts about fifteen active members and hosts a pair of annual rallies: one a camping event on the third weekend in June and the other a city ride on the last weekend in September.

Members range in age from around twenty to forty, and are drawn together by a passion for old Italian scooters. The club's literature states, "The PVSC isn't for the faint of heart. We don't toss newspapers, and we're not in bed by 9 p.m." Next time you're in Pittsburgh, stay up late with the PVSC.

The Checkered Demons Scooter Club at Fox Studio in New York

Checkered Demons, New York, New York

The Big Apple's most visible scooter club, the Checkered Demons, ride both Lambrettas and Vespas through NYC's often unforgiving streets. Founded in 1987, the club's focus is squarely on vintage scooters. The club has organized the Checkered Demons Rally every year since 1988, and has a fairly rigid structure, with such formalities as dues, regular meetings, and elected officers.

The Donne Veloci Scooter Club is an all-girl club in New York City

Down South

The Imperials, Atlanta, Georgia

Georgia is becoming a hotbed for scootering, as the city has seen a surge in scooter shops and clubs. The Imperials are one of the largest clubs in the south, boasting nearly thirty active members. Favoring a "general lack of snobbery," the club allows all scooters, new and old. (Most members, however, favor vintage machines.)

The club is a combined effort and was established in 2002 by combining two clubs, Motoretta Atlanta and Scootlanta. According to the club, the merger revved up the Atlanta scene and, "thus began the first of regular Wednesday night meetings of scooter riding, trivia playing, drunkenness, good-natured fighting, occasional nudity, and general jackassery. Each one of our members is a hardcore scooterist with a deep devotion and steadfast commitment to riding."

Scarlett Fever, Atlanta, Georgia

The sister club of the Imperials, Scarlett Fever is an all-girl group featuring around a dozen members. The club was formed in 2004 with the assistance of an original member from the well established Secret Servix Scooter Club, Chrissy Hyder. Scarlett Fever assists the Imperials in organizing Georgia's largest rally, Deliverance, each September. The club also organizes women to attend the annual all-girl rally, Love 'Em and Leave 'Em, which takes place in a different warm-weather location each winter.

Memphis Kings, Memphis, Tennessee

Over the years, Amerivespa rallies have been responsible for bolstering enthusiasm for scootering in their host regions. The Memphis Kings is a fine example of the impact the event can have. The club credits Amerivespa

ARTS & SOCIETY

Page 12E : Sunday, Oct. 20, 2002 : The Sun

Just Married

KATE DANA AND SEAN DETWILER

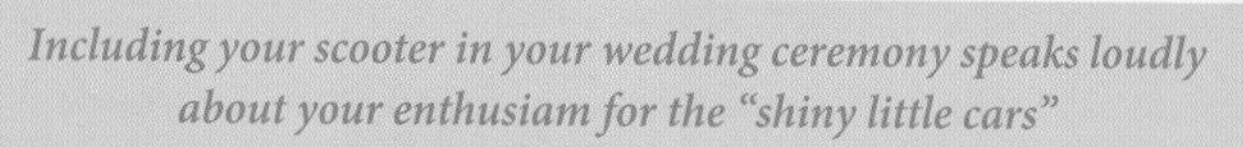

Including your scooter in your wedding ceremony speaks loudly about your enthusiam for the "shiny little cars"

with transforming it from a loose group of scooterists to a bona-fide club with its own annual rally, the Dead Elvis Rally held in August.

There are nearly thirty members, each with a fun nickname—some of these are "Wolfman," "The Dutchess," and "Snoop Bob." The Kings welcome scooters old and new and boast among their ranks the proprietor of a Vespa boutique dealership who also sells the venerable Stella model. With a solid core of enthusiastic riders and a famous rock-and-roll city to ride in, the Memphis Kings are only just beginning to stoke the scooter fire in Tennessee.

The Birmingham Scooter Syndicate of Alabama

Passengers allowed—Nomads Scooter Club

FOREST
PARK
BURGERS
freshly frozen
SOFT CREAM
got mullet?

Chapter 4

Runs, Rallies, Raids ... Mayhem

In 1949, sports journalist Renato Tassinari, who later founded the Vespa Club of Europe, did something really cool. He put out a call for Italy's *Vespisti* (Vespa riders) to gather en masse at the annual Milan Fair. And they did. Some 2,000 scooters took to the roads and went to the first scooter rally the world had ever seen.

The following decade, Italian scooters made a point of getting together on a regular basis, and their counterparts in neighboring countries followed suit. These early European gatherings were festive events that extolled the scooter's efficiency and technical charms. The big event was often gymkhana, a maneuverability contest based on equestrian competitions put on by British soldiers in colonial India.

Gymkhana proved too difficult and dangerous a pursuit for most rally-goers, although it remains a staple at modern events. Most scooterists wanted to ride their two wheels in the open air, to test their vehicle's limits, to see the world. For the next few years, European scooterists honed the art of the scooter raid, a cross-country excursion on the ever vibrating back of a scooter. In the early 1950s, an amphibious Vespa crossed the English Channel as other scooters took on mountains, vast cross-continental distances, and even around-the-world expeditions (executed in 1954 by an American named Dick Miller).

It didn't take long for the manufacturers to realize that scooter raids and rallies were ideal marketing opportunities, and they began organizing them. Ads hawking scooters in 1950s publications sung the glories of a two-wheeled road trip. "No hill too steep," touted a print ad for the Heinkel Tourist, "no road too rough, nowhere too far ..."

Of course scooter clubs were also instrumental in boosting the popularity of rallies. The first Vespa Club was founded in Saarbrucken, Germany, in 1951, and hundreds of scooter clubs followed in its wake, setting up shop all over the planet. By the middle of the decade, there were more than 50,000 card-carrying scooter club members worldwide; scooter newsletters were published, translated, and distributed in countries near and far.

Bank Holidays, Teddy Boys, and Mods

Scooter clubs soon proved ideal organizations to take charge of plans for rallies. In 1953, Club Lambretta of Great Britain became the first national scooter club to organize runs and rallies. The club got together the first Friday of every month, organized scooter runs into the

country, and made a steel badge for members to install on their legshields. But the club's events proved almost too popular: Innocenti sensed a marketing opportunity and pounced on it. As a result, the British Lambretta Owners Association (BLOA) came to be in the mid 1950s as an initiative of Innocenti and the company's British dealers.

By 1956, BLOA had pretty much superseded Club Lambretta, and every Lambretta buyer was automatically granted membership, supplied with a card and a badge at the time of purchase. Regional committees emerged to help organize massive national rallies. The Isle of Man Rally was first held in 1956 and went strong for two decades. The club sponsored international rallies as well, but they were never as well attended as the events held in the British Isles. BLOA became the Lambretta Club of Great Britain in the early 1960s, and later spawned similar clubs in other countries.

But before scooter rallies could become an international phenomenon, British scooter culture went into a state of semi-hibernation as big cars and motorcycles rose in popularity with the economic tide. In this same time frame, the British adolescent market emerged. Until the dawn of rock and roll in the 1950s, Brits had always gone straight from childhood to adulthood, strictly following the examples set by their elders—not anymore. Cheap, efficient scooters were easy to find at this juncture in British history, and became the transportation mode of choice for the teddy boys and girls, (a. k. a. "the teds"). Inspired by the lavish fashion of the Edwardian era, teds caused an uproar, riding their scooters, listening to American rock and roll, and generally inciting panic in the older generations with their unwillingness to follow tradition.

The rate of change within British youth culture quickened, and the teds gave way to mods and rockers in the early 1960s. The mods, outfitted with the latest in Italian fashion, saw their scooters as extensions of their meticulously crafted images. This, of course, was the complete opposite of the mods' rivals, the motorcycle-loving, greaser-inspired rockers. Both groups enjoyed the, "British working-class tradition of migrating to coastal resorts over Bank Holiday weekends," wrote Gareth Brown in his first-person account of the era, *Scooter Boys*. With swarms of scooters descending on the same destinations as herds of motorcycles, there was bound to be trouble. The violent clashes between mods and rockers in the mid 1960s have since become the stuff of legend.

The mods faded, as did British scooter culture, after the public became leery of them as the media fanned the flames with every mod-rocker incident. The hippies took over the U. K. youth culture, followed by the skins, the greasers, and the punks. In 1978, in the wake of the punk explosion and subsequent implosion, a mod revival started to take hold. The film version of The Who's *Quadrophenia* came out in 1979 and stoked mod nostalgia to a fever pitch. All of this meant that scooters—which were readily available for as little as forty pounds a pop—were back in style again and in a big way. Scooter shops opened, restorations were coveted, and scooters zipped all around jolly old England once again.

The renaissance of British mod culture led to the renaissance of British scooter runs. Packs of scooter-riding mods swarmed to coastal tourist destinations like Brighton and Scarborough once more. The Lambretta Club of Great Britain threw a members-only bash over the August Bank Holiday in 1979 and the rally ball really got rolling. Ensuing rallies were regional and informal, relying on word of mouth and little else, until the Lambretta Club of Great Britain organized a democratic process in 1982.

Thanks to better communication, British scooter rallies got bigger and bigger, culminating in the 1984 Isle of Wight Rally with 12,000 people in attendance. Tourist towns welcomed the scootering hordes because they spent plenty of cash on beer, food, and other rally necessities. But at the 1986 Isle of Wight event, riots broke out, the beer tent looted and burned, and responding firemen pelted with rocks. According to *Scooter Boys*, "Anarchy reigned supreme." The scootering community at this point realized the rallies had gotten too popular

"[Rallies] offered continual adventure, a newfound direction, and a ... sense of belonging ... . Scooter boys—and girls—wanted to travel, see other parts of the country, meet its people, party with others sharing their ideas, and live life to the full."
—Gareth Brown on early 1980s scooter rallies in the U. K. in his book, Scooter Boys

for their own good and intentionally scaled them back to get the focus back where it belonged: the scooter.

Since then, British rallies are smaller affairs; the largest typically attract about 1,000 scooter boys and girls. International rallies take place in France, Germany, the Netherlands, and other European locales, but those held in the U. K. remain the best attended of all.

And Now the Yanks ...

As was the case in Great Britain, manufacturer-sponsored scooter clubs were the first to organize rallies in North America. The Vespa Club of Boston put together runs and rallies in the 1950s, as did other regional clubs, but these events never garnered the popularity they had in the U. K. and the rest of Europe. Issues of *American Scooterist* from the 1950s rarely promoted organized events of any kind. But the manufacturers of the day used cross-country travel as a selling point: Cushman urged prospective buyers to "Discover New Frontiers" on one of their shiny scooters.

The popularity of scooter rallies in the U. S. was on the wane through most of the 1960s and 1970s, with most annual events surviving only a few years before going ka-put. Promotion was typically relegated to homemade flyers and word of mouth. Needless to say, there was no Internet to help spread the news.

But the 1980s saw the mod revival jump across the pond to the North America, catalyzed by the 1979 release of the movie *Quadrophenia*, and the American

mods mimicked their British forebearers not only in terms of their taste in fashion, but also their taste in scooters, and by that same token, scooter rallies. The North American rallies of the era were typically informally organized and tied to ska concerts that melted into all-night parties. The scooterists attending these rallies might well have talked differently than their counterparts, but almost everything else was the same. These Anglophiles cobbled together a lifestyle from British parts of all kinds; scooters were just one element among many.

However, a segment of the North American mods elevated the scooter up and above the rest of their accessories, and rallies took on a new form. Super Scoot, organized by Scooterville USA in Anaheim, was the big annual rally, a frenzied sixty-mile ride to Lake Elsinore for a weekend campout. Super Scoot peaked at scooter counts around 500, and then fell off the map by the end of the decade. Go Scoot Go in Berkeley carried on after the end of the Super Scoot era, as did a pair of long-running rallies north of the border: the Garden City Scooter Rally in Victoria, British Columbia, and the Niagara Scooter Rally in Ontario.

In 1986, the scene saw the first Scooter Rage, organized by the San Francisco chapter of the Secret Society. As the next generation of scooter kids re-ignited the rally fire in the 1990s, Rage was one of the few hold-overs from the previous decade. Scooter Rage is now the longest running, ongoing rally in the U. S., and one of the largest. Northward, in the Pacific Northwest, the inaugural Seattle Scooter Insanity was held in 1987.

But events that originated in the "Me Decade" are few and far between. It wasn't until the mid 1990s that the American rally scene really picked up steam again, thanks

Few souls are brave enough to gymkhana in the nude

in no small part to scooterists connecting on the Internet. The West Coast was the region primarily responsible for the rally's rebirth, with Rage, Insanity, and other events helping spawn such rallies as Orange Crush in Orange County and Kings Classic in San Francisco. Bay Area rallies inspired outsiders to start rallies in their own hometowns, and scooterists who attended these rallies were likewise inspired, until the rally calendar was jam-packed from January to December, and from coast to coast.

Gone are the days where American and Canadian scooter kids gawked at the European scooter magazines in stunned amazement. Now it is the Europeans who are taken aback by the fabulous restorations and customizations coming out of the North American scene, which has become a bona-fide "scooter place" with an identity and culture all its own.

What's What: Elements of a Scooter Rally

Scooter rallies are a unique phenomenon and getting to be more common than motorcycle rallies. There's more of a social aspect at play here. The lone biker is the prototypical image for motorcyclists, who crave freedom and independence; scooterists are more sociable almost to a fault. Typically traveling in groups, scooterists can talk while riding their machines, something that's very hard to do on a motorcycle because of the engine noise.

Beyond the shiny machinery and all the flirting and joking, a rally has many prototypical elements. Group rides are a big one. The organizers strive to put together the best rides their city—and its surroundings—has to offer. In Colorado, rides explore the Rockies, whereas every California rally is sure to explore the coast. No matter where you are, if you like to turn heads, there's nothing quite like a pack of several hundred scooters cruising the boulevard.

Gymkhana is another common rally component. Originally derived from contests between mounted British soldiers in India, scooter gymkhana naturally replaces the horse with a scooter, but continues to test the rider's ability to control their mount in the face of obstacles.

Another contest of skill often seen at American scooter rallies: the slow race. Using their balance, their wits, and their best ability to zigzag back in forth while moving forward as slow as humanly possible, contestants pilot their scooters to the finish line in hopes of being the absolutely last person to cross it. One of the few events in the entire sports world that is not about speed, rather the lack of it, a slow race is safer than gymkhana and therefore typically attracts a bigger, broader group of entrants.

Then there are the events where the scooters are not in motion: the parties, the mixers, the concerts, the poker games, and the like. Almost every scooter rally kicks off with some sort of mixer on Friday (or even Thursday), and the big event is a party with bands, food, free beer (provided by a sponsor brewery), and scads of scooterists raising hell. (Some

The slow race is one of the most popular gymkhana events

Burnin' rubber in Philadelphia

events have a dress-up night for the mods to show off their pretty clothes.)

Rallies typically fund themselves through the sales of raffle tickets, which are hawked throughout the weekend in anticipation of a drawing. The prizes vary and depend on the size of the rally, but larger events give away slick new scooters donated by manufacturers, as well as everything from footwear to gift certificates. But buyer beware: Many rallies in California are cash grabs, plain and simple. Any scooter rally worth its salt invests its raffle profits into next year's event.

There's also the "show and shine" event, when everybody at the rally lines up their scooters to be judged, typically offering many different categories of trophies. At Amerivespa, the official annual rally of the Vespa Club of America, the categories tend to be very straightforward and focus on quality restorations and well-kept original machines. Other rallies have a little fun with their trophies. At Mile High Mayhem, Denver's premier event, the categories change from year to year and are often as ridiculous as "Best Blue Scooter."

Event winners celebrating their victories

Always included are "Best Ratbike," "Best Un-Restored Bike," "Best Mod Bike," "Best Paint Job," "Best Color Scheme," and "Sportique Scooters Manager's Choice" for best in show.

For some scooter folks, earning accolades at rallies is a huge motivator. They prepare new machines and unveil them at a specific rally or event. Almost every rally will include a shiny, freshly built scooter that somebody just cannot wait to show off. And the bar is rising: the quality of the scooters at rallies gets higher each and every year.

Of course, camping rallies can't follow the same template as an urban event. There are no venues, no nightclubs, and no couches to crash upon when you're out in the wilderness. It's also hard to transport the trappings for gymkhana to a remote campsite. There is, however, usually a bonfire that entices a few wild-eyed scooterists to jump in it during the wee hours of the morning.

Scooters might be the starting point, but rallies are not about machines and machines only. As *Scoot! Quarterly* managing editor April Whitney puts it, "It's all about being around positive, excited people."

West Coast: Ongoing Rallies in California

Scooter Rage

The longest running rally in the U. S., Scooter Rage is the handiwork of the San Francisco chapter of the Secret Society, and—as the club's name suggests—the event's origins are something of a secret. For the first fourteen years, Scooter Rage was typically held the first of January, making it a shoe-in for the first rally of the year. But organizers got tired of freezing in the rain around Y2K and moved it to the third week of June.

As a prototype for several other ongoing city rallies, "We try to incorporate as much of the 'San Francisco experience' and typically hold events all around the city," says Mike Zorn, a longtime Secret Society member and Rage organizer. "We gave up on trying to throw the biggest rally years ago and have concentrated on making it all as fun as possible."

Fortunately gas stops are infrequent due to the scooters amazing gas mileage

Drawing about 200 bikes and 300 to 400 people, the schedule begins with a big party in the thick of the City by the Bay. "Our traditional Friday night beer bust kick-offs have become legendary," says Zorn, "and we try to keep our rides interesting and exciting." There are two rides on Saturday—a big ride into the mountains and a small ride around town—before the day show of gymkhana, a raffle, and other rally staples. The party Saturday night features DJs and, "the best bands we can find," says Zorn. On Sunday, a scavenger hunt or poker run, where the group is split into numerous teams, leads in to the wrap-up barbecue.

The Secret Society organizes all of the events, with a little help from friends near and far. "We have the advantage of having two great shops in town who have consistently supported us throughout the years," says Zorn, giving credos to First Kick Scooters and the San Francisco Scooter Centre. "We have also been lucky to work with a lot of the national shops as well as bike and equipment manufacturers, and we've given away everything from apple pies to brand new scooters at our raffles."

"We're not in it for the money, so we try to keep things like food, drinks, and shows free—or super cheap," he adds. "And when things seem like they are getting out of hand, they always seem to blow over smoothly."

Kings Classic

Held in San Francisco the second weekend every August, Kings Classic is the first event that many scooterists circle on their rally calendars. It's one of the definitive city rallies, and the rally that firmly established the prototype structure used by many younger events.

The first Kings Classic was not actually called Kings Classic. Organizers Josh Bluh, Sam Rogers, and Peter Gowdy dubbed the 1993 event Scooter Thing, bringing their collective background in concert promotions and classic car clubs together with a shared passion

"Anytime we get 500 scooters together, unless you're counting them, it looks like a million."
—San Francisco Scooter Centre's Barry Gwin

for vintage scooters. Scooter Thing was something of a pirate rally, promoted with a punk-rock attitude and legitimized by exactly zero permits from the city. The Rally Kings Scooter Club took the event over in 1994 and renamed it Kings Classic in honor of, well, themselves.

Barry Gwin, an organizer of Kings Classic (and owner of the San Francisco Scooter Centre), says it set the precedent for modern city rallies, in that it was the first annual rally to hold a raffle and the first to adopt a three-day calendar commonly favored by later rallies. Kings Classic begins with a ride Friday to a show, continues with a big party at a venue Saturday night, and wraps up Sunday with the major ride, which has traditionally ended with a "monster barbecue" in Golden Gate Park. "That formula pretty much turned into the formula for every three-day scooter rally in the United States," Gwin says.

Kings Classic attracts 300 to 500 scooters every year, and such numbers allow the rally to take over the city in a sense with head-turning group rides that run just about every light in the city. The initial ride on Friday night heads to a bonfire on the beach or a party at a club (or both). Saturday's ride takes the swarm to a day show, usually held indoors in a large nightclub, with the raffle, gymkhana, slow races, and judging for an awards ceremony.

As scooterists marry and multiply, Kings Classic has gotten increasingly family friendly: an inflatable castle for the kids to jump around in is another day-show staple. Not that the event's been stripped of its edge. One post-millennial day show also included a bunch of Britney Spears look-alikes wrestling in Jell-O. "It got a little bit naughty," says Gwin.

As of 2002, the resurrected Black Sheep Scooter Club (of which Gwin is a member) took over the day-to-day necessities of throwing Kings Classic. Organizers have shunned corporate sponsors, and instead fund the entire event off of the raffle and the financial support of the San Francisco Scooter Centre. "I don't think we've ever made money off of it," says Gwin, who says the task of throwing such a party gets tougher every year. "The older everyone gets and the more responsible everyone gets, the harder it gets to throw these massive rallies, because the liability is so high."

Orange Crush

Founded by holdovers from the 1980s Orange County scooter scene, the Hard Pack Scooter Club was at the forefront of the rally revival in the early 1990s. Motivated by fondly hazy memories of Super Scoot, the Hard Pack set out to bring back the days when a literal horde of scooterists would gather for a ride, and follow it up by partying into the wee hours of the morning. "That's exactly what we wanted," says founding member Charlie Yoon. "We wanted to see 500 scooters again."

After a few years organizing one-day events, the Hard Pack set its sights on something bigger, an event with, "a lot more riding," says Yoon. The result: Orange Crush I, a three-day party first thrown over Memorial Day weekend in 1995. Within a few years, it became the largest rally in southern California, attracting upwards of 300 bikes a year.

Friday night's meet-and-greet kicks off Crush, typically at a bar in Costa Mesa. The group meets Saturday morning in Newport Beach for a catered breakfast before embarking on a ride on the picturesque Pacific Coast Highway, typically fifty miles or more. Saturday night is the big club night, with bands and DJs and mod fashions galore. Sunday is the custom scooter show, with a raffle, gymkhana, and barbecue.

Because of corporate sponsors, "We're able to pull off more elaborate events," adds Yoon. "Every year, it seems we get a little more cash donations. We're trying to make sure everybody has a good time with as little expense as possible."

Cascadia: Ongoing Rallies in the Pacific Northwest

Seattle Scooter Insanity

First organized in 1987, Insanity is the Emerald City's longest running scooter rally. Victor Voris, owner of Seattle's Big People Scooters, helped put together the inaugural event held at the Mountaineers Club. Voris and company had been inspired by the Garden City Scooter Rally in Victoria, British Columbia. About 150 scooters swarmed the Mountaineers Club, northwest of the Space Needle, for the first Insanity. It was well established by the early 1990s, featuring a day show, raffle, slow races, and other rally staples. In recent years, the Vespa Club of Seattle has emerged as the prime organizer.

After a meet-and-greet the first Friday night of July, postmillennial rallies have included a breakfast in historic Pioneer Square Saturday morning, followed by a ride to the Experience Music Project at the Seattle Center for the day show. The site of more than 100 scooters around the base of the avant-garde museum (inspired by the shape of a smashed guitar) is a visual feast. Trophies are awarded for best restoration, the scooter that traveled farthest to be there (with winners often making the trek from the Bay Area), and others. One highlight from the years: The destruction of a Honda scooter that involved smashing it, dragging it, and crushing it with a van.

Garden City Scooter Rally

The continent's longest continuously running scooter rally is not in the U. S. but in Victoria, British Columbia, a Pacific Northwest scooter tradition since 1980. In the early years, Vancouver-based scooterists started taking the ferry west to Victoria for a springtime run around charming Vancouver Island.

Martin Wales helped organize the rally in the late 1980s and early 1990s from Vancouver, then moved to Victoria and found himself being the *only* organizer in

1996, 1997, and 1998. Overworked, Wales put a classified ad in the local paper urging local scooterists to unite at a local pizza joint, and the Capital City Scooter Club—the rally's organizers beginning in 1999—was born.

Since the mid 1990s, the Garden City Scooter Rally has consistently been the best-attended rally in the Pacific Northwest, drawing 100 to 120 bikes each year, rain or shine.

Positioned as the first rally in the Pacific Northwest, the schedule kicks off the Friday before Victoria Day (late May, for you Yankees) with a meet-up followed by a house party at a Capital City member's place into the wee hours of the night. Saturday features the weekend's longest ride. "There's a lot of parkland and farmland around Victoria," says Wales. "It's really beautiful, especially that time of year."

After returning to Victoria and meeting the ferry carrying the Vancouver contingent, the two groups merge and head out on another ride before reconvening for dinner and post-meal entertainment in the form of either a band or karaoke. "People get dressed up," says Wales. "All the mod suits come out."

Everybody meets on a cobblestone block for a parade of a ride down Government Street—the main tourist strip—before a ride. An awards ceremony and the raffle drawing cap the raffle at a Victoria nightclub Sunday night.

The High Country: Ongoing Rallies in the Rockies and the Southwest

Mile High Mayhem

Drawing as many as 280 machines, Denver's largest scooter rally is held over the last weekend of July. The first Mile High Mayhem took place in 1998, attracting people from both coasts and everywhere in between. It has since become a model for other rallies: Just as Mayhem drew inspiration from the experiences of Denver's scooterists

"Thus the Vespa came to be linked in my mind with transgression, sin, and even temptation ... the subtle seduction of faraway places."
—Italian writer and academic Umberto Eco in The Cult of Vespa

scoot moab 6

READ THIS! Thanks for coming to Scootmoab! We're in the desert, so please play safe. This can be a dangerous place. Drink lots of water and use sunscreen. Mind your riding ... it could take a couple hours for an ambulance to arrive if something happened. There are no gas stops on any of the rides. If you can't go 70 miles on a tankfull, put a gas can in a chase truck. Moab police radar on the canyon road. They also don't hesitate taking people to jail for drinking and driving or being a jerk.

A FEW RULES! **1)** Please mind the campground host and be nice to him. Fastest way to get us tossed out of here is to flip the host some shit. **2)** Go slow in the parking lot. **3)** No starting your scooters after dark. Push it out to the street. **4)** Have fun at night, but please be respectful and keep the noise down. **5)** When we ride through the gate into Arches NP, go slow, ride to the back of the parking lot and shut off your engine. Otherwise, we'll have rangers follow us the whole ride.

FRIDAY

Deadhorse Point ride .... 3:00 p.m.

70 miles. No gas stops. No slow bikes. Look at the ride map on the right. We will ride down (southwest) highway 128. Stop at the bottom at the intersection of 128 and 191. Go north, then head west on 313 to Deadhorse Point. Amazing ride! $3 to get into the park. If you think you'll need gas, put a gas can in the chase truck.

Dinner .... 6:30 p.m.

SATURDAY

Coffee, KrispyKremes and Bagels .... 8:00 a.m.

Arches ride .... 11:00 a.m.

60 miles. Gas stop in Moab at the end of the ride. Look at the ride map on the right. We will ride down (southwest) highway 128. Stop at the bottom at the intersection of 128 and 191. $5 admission ($10 if you're riding 2-up. WTF?) After our stop at the ranger station, we will ride to the Windows section for a stop (we may split into two groups from here, depending on interest). After the 30-45 minute stop, we ride to Devil's Garden for another stop. Then ride back into Moab for a late lunch. Choose your restaurant! We don't have reservations anywhere. Main place will be Tax's, but it could be very busy.

Moab Olympics .... 4:00 p.m.

Photo portraits on scooters .... Around sunset

Raffle .... 7:00 p.m.

SUNDAY

Coffee, KrispyKremes and Bagels .... 8:00 a.m.

Ride to Castle Valley .... 9:00 a.m.

Low-key ride northeast on highway 128, then take a right into Castle Valley. It gets a little cold, so bring a jacket. No gas stops. About a 45 mile loop. Great way to end the rally!

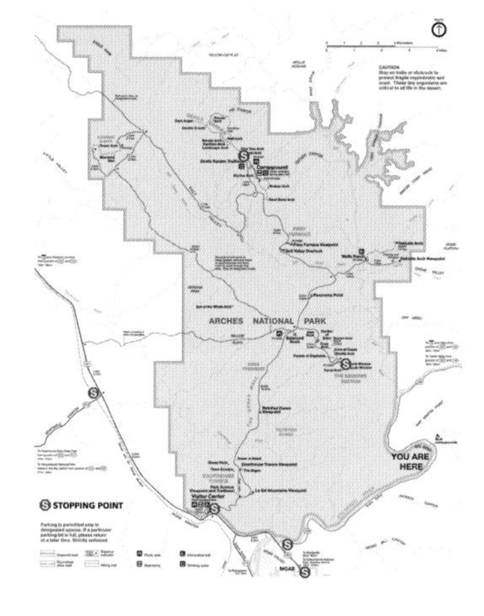

Attendees to Scoot Moab 6 received a laminated map detailing rides and schedules

in San Francisco, visitors from the Bay Area took lessons home from the first rally. New York's Gotham was inspired by (and structured after) Mayhem as well.

The fullness of the schedule makes Mile High Mayhem unique. It's wall-to-wall events from Thursday night to Monday afternoon. Things warm up Thursday evening with a big mixer at a pub. Friday morning is the shopping ride, complete with a coupon book donated by local merchants. The alternate activity is the "gentlemen's lunch" sponsored by the ACE Scooter Club at a topless club. Come Friday night, the girls and boys meet in separate locations to ride to a show with free microbrew and pizza. On Saturday, hundreds of scooterists rise and shine for a ride into the Rocky Mountains, the destination being a scenic picnic spot for a catered lunch. The post-picnic time slot is open, giving the mods plenty of time to dress themselves up for the blowout on Saturday night, featuring bands at a concert venue. The late night/early morning tradition is a massive after-hours warehouse party.

For better or worse, everyone meets at a coffeehouse Sunday morning for a "parks and parkways" ride through Denver that ends at Sportique Scooters just west of downtown. After the morning-after-haze wears off, Sunday afternoon sees one of the hairiest gymkhana events held anywhere in the U. S. Scooters have been known to get seven feet of air, and sometimes get smashed to smithereens. A series of slow races follow gymkhana; during both events, Sportique's crew judge the scooters for the rally's dozen awards, as well as the Erickson Scott Memorial Trophy, dedicated to a dearly departed friend of Denver's scootering scene. The trophies are awarded at an evening ceremony at a Denver nightclub, and the rally's raffle follows.

Sunday night is reserved for an all-out bout of karaoke at a local bar, followed by an all-night poker game at a private residence. By 7 a.m. Monday morning, whether your chips are up or down, you're usually too tired to sleep.

With no registration fee of any kind, Mayhem is entirely sponsor-supported and the raffle is the largest of any scooter rally in the country; past Mayhems have had as many as four scooter giveaways. There is no profit taking, one year's raffle-ticket sales fund the next year's rally.

Scoot Moab

Organizer Dana Wilson points to two key factors that underpinned the first Moab rally: The success of Mile High Mayhem in Denver and the emergence of the Internet as a hub for the scooter community. After he went to a few Mayhems, he got online and started organizing a camping rally in the Utah desert.

Held in the spring of 2000, the first Moab rally wasn't actually anywhere near Moab—it took place on the other side of Utah in Zion National Park. The park

"It has less to do with the bikes and more to do with the people. They're kind of off. There's something wrong with them. But people will travel across the country with a beat-up, crappy bike, not because they want to show it off, but because they want to hang out with the people."

—Sportique's Adam Baker on the allure of the scooter rally

Keep a watchful eye on your moped when attending Camp Scoot

service was getting ready to permanently close the main road through Zion to all traffic, aside from official park buses, and Wilson realized the window of opportunity for a scooter rally in the park was quickly closing. He organized the event for the last weekend in April and attendance far exceeded expectations with sixty-eight scooters and about 100 people, many of them from Denver. Two weeks after the 2000 rally, Zion administration closed the main park road to scooters, not to mention cars and motorcycles, so the hunt was on for a permanent location. Wilson picked a group campsite on the Colorado River between Moab and Arches National Park—in part because it was more convenient to the Denver contingent.

Scoot Moab's schedule includes one or two rides a day, emphasizing the desert in all of its rock-and-sand glory. The routes run through Arches National Park, the nearby La Sal Mountains, and other blissfully scenic areas. In sarcastic honor of the 2002 Winter Olympics in Salt Lake City, the agenda also includes its own tongue-in-cheek Olympic Games. Past events have involved shot-putting a Lambretta cylinder, discus-tossing a scooter hub, and breaking a piñata while riding a scooter to empty it of the adult novelties and mini liquor bottles within.

Scoot Moab relies on a nominal registration fee and raffle ticket sales to pay for the food and the group campsite. Wilson never quite breaks even, but to him, safety is more of a priority than profitability. "The desert's a dangerous place and I want people to be careful," he says. "I've always been pretty big on the safety, and not letting people drink and ride."

Camp Scoot

Starting in 1995, Albuquerque's Nomads started the annual tradition of rallying the local troops for a run into the nearby Jemez Mountains. The Nomads' event lasted until 2000 (for the last two years it changed destinations to the Sandia Mountains and was renamed Dead Man's Run), and then faded into the scooter rally

A beautiful setting for Mayhem 2003

history books. But in 2003—sparked by the opening of a new scooter shop in Albuquerque, New Urban Transport—the Atomic Scooter Club emerged and resurrected the annual August pilgrimage to the Jemez Mountains.

Rally-goers meet at New Urban Transport and brave the 100-mile ride to a campsite in the Jemez Mountains. After tents are pitched and camp is prepped, the afternoon's prime event is a short ride through scenic canyonlands. Saturday's ride takes the group to Los Alamos, a fifty-mile ride past the Valles Caldera, an enormous crater that was the handiwork of a now-extinct volcano. After lunch in Los Alamos, the group heads back to the site for gymkhana and a game called "beer bungee." Explains organizer Sean Campbell, "You tie a big fat bungee cord around a tree and tie the other end to a person. You make them run against the bungee cord, over a slickened tarp, to try and grab a beer off the table."

Organizers charge a "fairly steep rally fee," says Campbell, but it covers the campsite and all meals. The event usually attracts about fifty bikes in all and is known for its pranks. One event saw the sun rise over a moped flying full-mast on the campsite's resident flagpole.

Riding the open roads in the desert Southwest draws scooterists from all over to these rallies

Tucson/Nogales Fall Classic

One of the longest-running camping rallies in the country, the Tucson/Nogales Fall Classic started in 1988 with a few scooterists taking a costumed day ride on Halloween to Mission San Xavier del Bac. It grew into an overnight trip the next year when Scott Dion and Tim O'Brien grabbed the reins and scouted out the U. S./Mexico border area near Nogales, about 127 miles south of Tucson. They found a gem of a campsite with a central fire pit and the rally sprouted from there, attracting about 100 bikes a year—90 percent of them from outside of Arizona.

Most years, rally-goers meet the Friday of the first weekend in November at the historic, hip Hotel Congress in downtown Tucson and party at its resident rock club. ("There was one year they didn't allow us back," says organizer Matthew Noli.) By noon Saturday, the horde congregates and takes off on the journey to the campsite, with the last scooters trickling in just before sundown. People pitch their tents on the outskirts, and the party is centered around the bonfire.

"As things get crazier and crazier, the infamous fire jumping and running around naked starts going on," says Noli, a tradition launched by the Pharaohs in

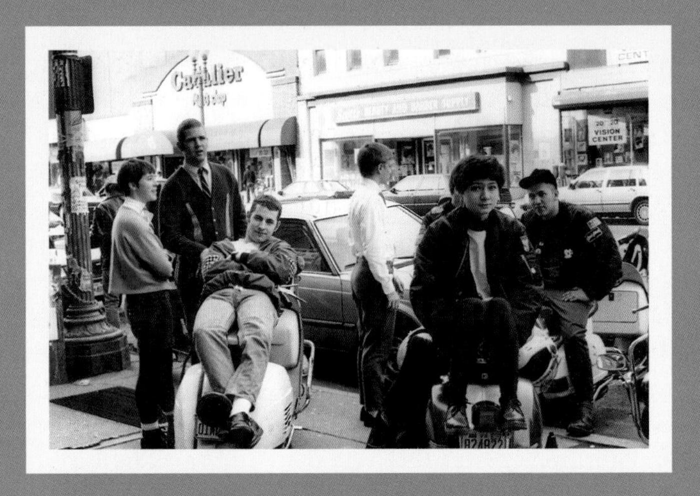

"It's loud, it's rowdy, it's a whole lot of drinking, it's a whole lot of everything."
—Tundra Schmuck Ian Whalley, describing the Niagara Scooter Rally

1997. "They were really stoked, because they didn't have to wear helmets, they could carry their firearms out in the open, and they could drink and smoke in restaurants without being hassled. At the campsite, they were really wild … jumping a ten-foot bonfire on a makeshift ramp. It was quite a show."

On Sunday, "Everybody crawls out of their tent or their hole in the ground," Noli continues, "and rides back north to a park in Tucson for a barbecue and the raffle giveaway." The event is supported by the raffle—donations from merchants near and far—and rally-pack sales.

High Rollers Scooter Rally

Held in Las Vegas since 1999 in mid-February, the High Roller Scooter Rally has the reputation of being the least organized, most chaotic scooter rally in the country. And it's also one of the biggest, attracting upwards of 500 bikes and 1,000 people in years when strong British contingents make the trip. (In 2003, about half of the 900 rally-goers hailed from the U. K.)

With endless ribbons of neon, free booze, and the rampant vice, Vegas is something of an unruly rally setting, and the big events have been known to splinter into cliques cruising the strip on their own. But there is hope. "I'm really pushing to get the Vegas rally more organized," says Boston-based Eric Porter.

The weekend typically kicks off with a "dance" Friday night. Saturday morning begins with the show and shine, followed by a big group ride Saturday morning. (Hoover Dam and Red Rock Canyon have been past destinations.) The Saturday night blowout features bands and DJs; rally-goers generally carry on all night long—and then some. Sunday includes a pair of optional rides leading into a barbecue with gymkhana and an awards ceremony. In between events, attendees have been known to get hitched at one of Sin City's quickie wedding chapels.

Martha Reeves poses with attendees at the 1992 Ambassador's Motown Scooter Rally

Rally-goers have traditionally stayed at the Gold Spike Hotel & Casino in downtown Las Vegas. "The Gold Spike is cheap, and scooterists are cheap, so there's sort of a nice marriage there," notes Porter.

"Vegas is just the backdrop for the rally," adds Porter. "It's not about gambling. For me, it's an adventure. We crate up our scooters, ship them to Vegas, unpack them, ride them around for three days, re-crate them, and ship them back to Boston."

Middle America:
Ongoing Rallies in the Midwest

WKRP in Cincinnati

Organized by the Ten Year Lates, a Cincinnati scooter club, this rally was first put on by a handful of clubmates in 2001, dubbed "A Decade Late" and attracting a mere half-dozen bikes. Just three years later, more than 150 scooterists showed up to pay respects to Dr. Johnny Fever and Venus Flytrap at the 2004 rally, making it one of the biggest events of its kind in the Midwest.

Centered on the Comet, a hip Northside bar and music venue, WKRP is held over the last weekend in March and kicks off with a low-key, brewery-sponsored party on Friday night. Breakfast at one of the Lates' houses eases the group into Saturday, which also features a ride from downtown Cincinnati out to its rural surroundings, then back into town for an afternoon barbecue. After a round of gymkhana, the rally-goers congregate once again at the Comet for bands, karaoke, and a raffle.

"The biggest selling point for our rally is that everything's free," says organizer Casey Beagle, noting that "everything" includes everything from T-shirts to breakfast to beer. "That, and we have some really good rides around here."

Slaughterhouse

On their trip back from the Niagara rally in 1995, Moe Balazs and his wife Kristin envisioned a scooter rally in their hometown of Chicago. Once home, they started the Second to Last Scooter Club with the express intent of organizing a rally, and did just that a few months later with Slaughterhouse I, a camping rally on the outskirts of the Windy City's outermost suburbs.

It was a success, including a ride downtown and a lakefront barbecue. "We made 100 patches and sold them all," says Balazs. But ensuing Slaughterhouses didn't necessarily follow the same recipe. "We've done camping rallies and we've done city rallies. We kind of alternate."

The urban Slaughterhouses have typically centered on the Hideout, a bar favored by scooterists on Chicago's Near North Side. "It's in an area that's all industrial," says Balazs. "During the evening, the streets

Riding with pride in Philly

are wide open, so it's a great place to have a rally." The camping Slaughterhouses, after the first one, tended not to include rides to downtown Chicago, because of the tricky logistics. Both have been traditionally held over Labor Day weekend and draw about 100 bikes.

One camping Slaughterhouse was on the property of a biker bar called E. J. Karz, about seventy miles southwest of downtown Chicago. "By Sunday morning, they were down to Coors Light and schnapps," Balazs recalls. "They had nothing to serve their biker guests because we totally depleted their bar."

After Slaughterhouse X in 2004, Second to Last Scooter Club started working with other Chicago-area clubs to organize future Slaughterhouses. Says Balazs, "There's a lot of younger people who are gung-ho here, so we're hoping they'll pick it up."

Back East: Ongoing Rallies

Gotham

The 2002 Mile High Mayhem set in motion a pair of NYC scooterists, Stian Nilsen and Niabi Caldwell, to organize their own event in the Big Apple. "We were loving how well-organized Mayhem was—we were inspired," remembers Nilsen. "Niabi and I took the ball and ran with it."

Held the first weekend of May 2003, the first Gotham Rally sported the tagline, "The Rally that Never Sleeps," and that it did. There were events scheduled for 4 a.m. "We had a ride that watched the sun come up," says Nilsen. More than 100 scooters were on hand for the festivities. Gotham II ("The Rally You Can't Refuse") had a *Godfather* theme, and Gotham III ("Notorious N.Y.C.") a hip-hop one. While Gotham organizers don't stick with a fixed schedule year after year, there are typically rides through Manhattan (including Central Park), Brooklyn

Rally Patches

The flight jacket has long been the scooterist's outerwear of choice. Cheap, readily available at military surplus stores, and resilient enough to withstand the elements, the flight jacket's primary use is as a canvas, plastered not with paint but with patches from rallies near and far. (Another common trick: Sewing a Newcastle bar towel onto the sleeve to wipe the rain from one's seat.)

The scooter patch is one of the staples of a rally pack, and it gives rally organizers a chance to get creative and come up with a logo that captures the spirit of the event. Over the years, these little scraps of fabric have become miniature works of art, coming in all shapes and colors, with logos ranging from simple and subtle to dirty and devilish. The Internet has emerged as a virtual gallery for rally patches—check out the patch archive at Scoot.net and the vaults at www.scooterpatcharchive.com.

District
69
S.C.

Gordy
MOTOWN RALLY
AUG. 27-29
1993
DARLING, I HUM OUR SONG
(Holland, Dozier, Holland)
MARTHA & THE VANDELLAS

Cannonball Run
Virginia Beach to Santa Monica
13 Days · 13 States
11 Riders · 3400 Miles

Old Bastards Scooter Club
OBSC
Life Begins At 24

TORQUAY
RUN
OCT

Whitley
Bay

BOSTON STRANGLERS
WICKED PISSAH RALLY
JUNE
SOUTH EAST
RAIDERS S.C.
AUGUST 1992
4TH ANNUAL BASH

RADUNO NAZ. LAMBRETTA
Lambretta
FIRENZE 28/30 5 93

LONE STAR ROUNDUP
LONE
STAR
Republic Of Texas Scooter Rally
Austin, Texas

BLACK COUNTRY
DELTAS S.C.
GREAT YARMOUTH RALLY
WHITSUN 1983

INCRIMINATORS
illud non fecimus
CHAPELHILLNC

Super
Custom Show
14TH NOV 93
DUNSTABLE

Washington, D.C.
checkered
demon
SCOOTER CLUB
NY/NJ
THE RETURN OF
THE
BIG WET ONE
GO! SCOOT GO!
CHECKERED
DEMONS
S/C
6th ANNUAL RALLY
Scooter
NEW YORK 92
Class A
S.C.

THE ALL-NIGHTER CLUB OF GREAT BRITAIN
19
86
MEMBER

GT. YARMOUTH
SCOOTER
RALLY 86
1986
MARCH 28·29·30·31

LONG ISLAND RUN
AUG. 4-6, '89

TAMPA TWO STROKE SCOOTER CLUB
RALLY
2004
SLEEP AWAY CAMP

VESPA CLUB D'ITALIA

D.C. Scooter Run
June 8-10 1990

Brigham's
Bees
Scooter
Club

SKOOT
AND
DESTROY
2strokeBuzz.com/racing

TORONTO '97
SCOOTER RALLY

Bacchus Raucous

BELGIAN RUNS
94

TRAVELLERS
SHEFFIELD
SCOOTER
S T
S C
CLUB

www.iscootny.com

rally in the fort
'03

THE BLAIR
SCOOTER
PROJECT
JUNE 2000
PVSC

(including Coney Island), and Queens, as well as a day show followed by a big blowout party on Saturday night.

After leading the charge for Gotham I, Nilsen and Caldwell turned organizing duties for subsequent rallies over to a committee composed of members of most of the city's scooter clubs. "It's like hosting a party," says Nilsen. "It's a lot of work—and a lot of fun. Besides being a good excuse to drink with your friends," he adds, "you get to see people coming together because of their love for these kooky little bikes."

Niagara Scooter Rally

Niagara became an annual north-of-the-border tradition in 1985, thanks to, "a bunch of Toronto folks who wanted to head down somewhere and party," says longtime organizer Ian Whalley, who describes a pattern that emerged in the early years: "Go to a campground, get kicked out. Go the next year, go to another campground, get kicked out."

What was fun for a few years became "an absolute mess" by the early 1990s, adds Whalley. "You couldn't find a campground anywhere in Greater Niagara we hadn't been kicked out of." By 1992, Niagara's legendary debauchery was endangering the event, especially after the group got kicked out of a place called Yogi Bear's Family Campground and couldn't find a campsite anywhere for the second night.

Whalley took the rally over the next year and moved it to the U. S. in order to prevent a few troublemakers with criminal records from attending. About 130 people showed up and Whalley continued to throw the party in upstate New York before connecting with the folks at the Welland County Motorcycle Club in Ontario in 1997. The rally has been held at their clubhouse ever since, and

Whalley says it's a perfect match. "They were pretty receptive to massive bonfires and drinking starting on Friday and stopping on Sunday," says Whalley. "Short of destruction, there isn't much we can do to get kicked out."

Held in late May over Canada's Victoria Day weekend, the Niagara Rally is one of the most storied camping rallies on the continent, drawing 150 scooters and 400 people a year. Between what Whalley describes as "a whole lot of blur," the rally is a three-day party punctuated by such storied traditions as a ride to Niagara Falls (and a stop at a strip club on the way back), a U. S. vs. Canada tug-of-war, and "mud drags"—involving scooters dragging a person sitting in a kiddie pool over a muddy field. "That can get pretty hairy," says Whalley. So can the traditional 2 AM naked rides.

Canadian Whalley founded the club behind the event, the Tundra Schmucks, with "honorary Canadian" Rob Stuhr in 1993. ("This American guy wanted to fight and called us tundra schmucks," Whalley says of the name.) After operating for a decade as a duo, the Schmucks expanded their ranks to a quintet in 2003.

Location TBA Rallies

All Girls Scooter Rally

The first All Girls Scooter Rally (AGSR) was held in October 2003 in Palm Springs, California, with the intent of going to a different location every year. The site of the second one—dubbed "Love 'Em and Leave 'Em"—was sunny Hollywood, Florida. The attendance doubled to forty scooters and fifty-five girls for AGSR II.

"Each girl that attends the rally is an organizer, and everyone chips in to make this the best rally it can be," says Missi Kroge, one of the minds behind AGSR. The major sponsors include many girls-only clubs like Secret Servix, Scarlett Fever, Baltimore Bombshells, Donne Veloci, and the Instigators.

"Since the location of the rally changes each year, the

scooter shop in the chosen location usually is our biggest sponsor," Kroge says. "In Palm Springs, Scooter Parts Direct was a major donator. In Miami, Casa Lambretta went above and beyond the average sponsor."

Events depend on the city: The Florida rally included a casino cruise into international waters, a craft fair, a ride into South Beach, and a show in the Miami Design District featuring a bunch of different bands.

"This rally encourages scooter girls to get to know each other and bond," notes Kroge. "Its intent is to let girls know that rallies aren't just for boys anymore."

Amerivespa

The annual late June rally of the Vespa Club of America (VCOA), Amerivespa was first organized in 1993 in Springfield, Missouri. It was held again in Springfield in 1994, and for two years each in Manitou Springs, Colorado; Knoxville, Tennessee; and San Diego, California, before moving to a different city every year when it came to Oklahoma City, Oklahoma, in 2001. From 2002 to 2005, the next four Amerivespas were in Portland, Oregon; Madison, Georgia (where the VCOA teamed with a microcar rally for a one-of-a-kind event); Salt Lake City, Utah; and Cleveland, Ohio.

While VCOA President J. D. Merryweather wants to look to more East Coast cities in the near future, he also has a vision of merging Amerivespa with Mile High Mayhem in Denver sometime, the result being "the biggest scooter rally ever."

Amerivespa attracts close to 300 scooters a year, and, not surprisingly, most (but not all) of them are Vespas. Merryweather says attendees are a mix of diehards who come year after year and local scooterists. "It changes year after year, depending on what city we're in," he says. The schedule follows the standard weekend rally formula, with a meet-and-greet Friday night, a show and gymkhana competition Saturday, and rides throughout.

Many rallies are "very exclusive," Merryweather adds, touting Amerivespa as an event open to young and old, even families. "Some people peg it as an old fogey rally, and it's really not," he says. "But it is a rally that acts as a really good bridge between young scooterists and older scooterists. I think you're starting to see a lot of the younger generation wanting the knowledge of the older scooterists."

65

Eintracht
CHICAGO

Coors
Coors

A Few Great Scooter Sites

Scoot.net
ScooterBBS.com
2strokebuzz.com
2-Stroke Smoke (yahoo groups)
scooterhelp.com
vespa.org
vespaclubusa.org
scooter.com
scootersociety.com
lambrettagirl.com
honestvaclavs.com
easthillcd.com
nightryders.tk

A Few Great Scooter Shops

Arizona
Scoot Over
Hot sun, cool scooters, and warm folks
4534 E. Broadway Blvd., Tucson
520 323 9090

California
Motorsports Scooters
Scoots, accessories, parts and performance
4225 30th St., San Diego
619 280 1718

West Coast Lambretta Works
Vince Mross and crew are tops in Lambretta
6244 University Ave., San Diego
619 229 0201

NOHO Scooters
Kymco and TN'G shop—vintage too!
5144 N. Vineland Ave., North Hollywood
818 761 3647

Moto Paradiso
Modern and vintage Vespas
707 Anacapa St., Santa Barbara
805 564 moto

Scootershop
Mick is old school and his shop rocks
1043 W. Collins Ave., Orange
714 289 1163

First Kick Scooters
New and vintage sales and parts
275 8th St., San Francisco
415 861 6100

San Francisco Scooter Center
Barry Gwinn's fantastic shop
127 10th St., San Francisco
415 558 9854

Bullet Proof Scooters
Josh and Eric paint their butts off
San Francisco
805 481 0599

Colorado
Sportique Scooters
The original shop
3211 Pecos St., Denver
303 477 8614

Sportique Scooters / Kymco of Boulder
Peoples in the republic of Boulder
2506 Spruce St., Boulder
303 402 1700

Pikes Peak Sportique
Focus on a family friendly scooter shop
431 E. Pikes Peak Ave., Colorado Springs
719 442 0048

Connecticut
Scooters Centrale
The center of the scooter universe in CT
161 Woodford Ave., Plainville
860 747 2552

Florida
Casa Lambretta USA
An offshoot of Casa Lambretta in Italy
2403 NE 2nd Ave. #B, Miami
866 lambretta

Moped Hospital
One of the top tuners of modern scooters
601 Truman Ave., Key West
866 296 1625

200cc Inc.
Scooter gear for those on the GO!
1636 Hendricks Ave., Jacksonville
904 306 9500

Illinois
Scooterworks USA
The top mail-order house in the USA
5410 N. Damen, Chicago
773 271 4242

Indiana
Speed City Cycle
Awesome custom work
3464 W. 16th St., Indianapolis
317 917 3211

Kansas
Scooter World!
A really cool shop in Overland Park
7325 W. 79th, Overland Park
913 649 4900

Kentucky
SoHo Scooters
Right outside of Cincy
625 Monmouth St., Newport
859 431 ROLL

Louisiana
Vespa New Orleans
A good little shop in the Big Easy
3540 Toulouse St., New Orleans
504 483 6776

Maryland
Moto Strada
Mark and crew have done it right for years
9918 C York Rd., Cockeysville
410 666 8377

Minnesota
Scooterville
One of the top Stella shops in the USA
650 25th Ave. SE, Minneapolis
612 331 7266

Missouri
Extreme Toy Store
Scooters, carts, and go-peds
3217 Lemay Ferry Rd., Saint Louis
314 892 4000

New Mexico
New Urban Transport
Julie and Nick give great scooter
1800 Central Ave. SE, Albuquerque
505 247 2698

Centaur Cycles
Richard has been at it for a quarter decade!
3101 Jemez Rd., Santa Fe
505 471 5481

New Jersey
Scooters Originali
Gene Merideth's shop specializes in Vespas
5 Lawrence St. #17, Bloomfield
973 743 6060

New York
Brooklynbretta
Lambrettas and Vespa service
644 Sackett St., Brooklyn
718 422 0556

North Carolina
Mountain Scoot Adventures
Scooter rentals in heaven
98 N. Lexington Ave., Asheville
828 252 9696

Ohio
Pride of Cleveland Scooters
Phil Watters' pride and joy
2078 W. 25th St., Cleveland
216 737 0700

Supersonic Scooters
Scoot Smallwood is a fantastic tuner!
1386 Fields Ave., Columbus
614 299 8480

Metro Scooter
Cincy's Stella hookup
3700 Montgomery Rd., Cincinnati
513 631 6637

Oklahoma
Atomic Brown Scooter Shop
The Stella place up in the OKC
4415 North Western Ave., Oklahoma City
405 605 3789

Oregon
P-Town Scooters
An old school shop in the heart of Portland
3347 SE Division St., Portland
503 241 4745

Pennsylvania
Skytop Scooters
The Rover shop with great scooters!
1045 Sarah St., Philadelphia
215 426 BRIT

Philadelphia Scooters
Yo' Philly's scoot shop
1737 E. Passyunk Ave., Philadelphia
215 336 8255

Rhode Island
Javaspeed Scooters
Woody will fix you a latte then tear
into your Series 3
1284 N. Main St., Providence
401 270 9485

Texas
Garners Classic Scooters
Randy was scooters when scooters wasn't cool
1414 Southern Blvd., Cleburne
817 645 3478

American Scooter Center
Tons of parts, projects, and pristine scoots
13804 Dragline, Austin
800 97 vespa

Utah
Scooter Lounge
Dave Hurtado is the man
824 S State St., Provo
801 434 4536

Virginia
SCOMO
One of the best mail-order shops in the USA
217 W. 7th St., Richmond
877 scoot 25

Washington
Big People's Scooters
Victor's old school shop rules
5951 Airport Way South, Seattle
206 763 0160

Interbay Scooters
Real good folks
1805 15th Ave. West, Seattle
206 284 9084

Wisconsin
Scooter Therapy
Take one scooter and call in the morning
9 N. Ingersol, Madison
608 255 1520

Canada
Motoretta
Vespas from yesterday and today
554 College St., Toronto, ON
416 925 1818

Scooter MD Services
Dennis and crew really know their stuff
368 West 1st Ave., Vancouver, BC
604 879 9501

Index

50cc, 46, 54, 55
ACE, 99, 100-101
Aermacchi, 26
All Girls Scooter Rally, 135-136
Allstate, 30-32, 34, 38
Ambassador's Motown Scooter Rally, 129
American Scooterist, 43, 114
Amerivespa, 46, 60, 88, 98, 101, 103, 107, 117, 136
Aprilia, 43-45, 47, 50, 51, 53, 54, 56-60
ASRA (American Scooter Racing Association), 72, 84-86, 96
Atomic, 101, 127
Audrey Hepburn, 28, 31-32
Auto-Glide, 15, 17
Autoped, 13
Bajaj, 42, 47, 53, 61
"Barn Find", 33
Basile, Ryan, 63, 88
Birmingham Scooter Syndicate, 109
Blue Smoke, 100
Bottle Rocket, 92, 99
Brigham's Bees, 98-99
BSA, 27, 83
Burgundy Topz, 94-95
Camp Scoot, 101, 126
Checkered Demons, 104, 106-107
Choppers, 70
Commuter Scooters, 47, 56-57
Constructa-scoot, 16
Continental Kings 103
Cooper Combat Motor Scooter, 17
Cooper Motors, 17
Culture, 27, 81-91, 112-113, 116
Cushman Motor Works, 14-31, 34, 35, 40, 54, 65, 88, 114
Customs, 69-75
CVT (constantly variable transmissions), 16, 54, 74
D'Ascanio, Corradino, 21-22
Derbi, 44, 53, 58, 59, 76
Eastern Scooter Racing Association, 85
Electric Vehicles, 48, 51
EPA, 48, 58, 59
First Kick Scooters, 41
Formocino, 8, 24, 26, 27
Fort Collins Scooter Gang, 100-101
Fuel Cell, 51
Fuji, 26, 27, 33, 73
Garden City Scooter Rally, 116, 122-123
Gas Conservation, 36-37
Go-Ped, 13
Gotham, 124, 131, 134
Gymkhana, 111-116, 121, 124
Hard Pack, 96-97, 121
Harley-Davidson, 16, 29-30, 46
Hell's Belles, 97
High Rollers Scooter Weekend, 95, 129
Hill City United, 104
Honda, 28, 39, 40, 42, 46, 56, 60, 122
Hybrid Vehicles, 51
Illustrated Motorscooters Buyer's Guide, 9, 15, 29, 81
Imperials, The, 107
Indian, 16, 29, 30
Innocenti, 19, 35-37, 112
ISO, 27
Italjet, 42-43, 53, 56, 58
Jedi Knights, 92-93
King Tut Putt, 93-94
Kings Classic, 94-95, 116, 119-121
Kroge, Missi, 86-87, 135-136
Kymco, 44, 50, 53, 54, 59, 60, 89
Lahmers, Chelsea, 87
Lambretta, is everywhere
Lifestyle, 45, 81
Lopez, Darrin, 88
Los Gatos Gordos, 97-98
Ludwig, Bryce, 89
Malaguti, 44, 53, 56, 57, 58, 59
Maxi Scooters, 46, 47, 50, 60, 62
McCaleb, Philip, 41, 47
Memphis Kings, 107, 109
Mid-size Scooters, 58
Mild Customs, 69-70
Mile High Mayhem, 86, 94, 99, 100, 102-103, 117, 123, 124, 131, 136
Minnesota Maxis, 59, 104
Mitsubishi, 26, 27, 32, 73
Mod Scooters, 72-73

Modern Scooters, 53-62
Mods, 72, 81-86, 111-115
Mods and Knockers, 99
Moto Guzzi, 26
Moto Rumi, 26
Motobi, 26
Motoped, 13
MV Agusta, 27
Niagara Scooter Rally, 96, 115, 128, 130, 134-135
Nomads, 101, 109, 126
Oppressors, The, 103
Orange Crush, 97, 116, 121
P Series, 38-39
Parilla, 27
Parker, Waid "Scooter Daddy", 88-89
Patches, 132-133
Peak, 101-102
Pharoahs, 93-94
Piaggio, 19-29, 31, 34, 39, 42, 44, 47, 57-59, 60, 66, 79
Pittsburgh Vintage, 106
Pride of Cleveland, 85, 102
Quadrophenia, 72, 83, 84, 101, 113, 114
Racing, 84-86
Racing Scooters, 57-59
Rallies, 116-118
Rally Kings, 94
Ratbikes, 71-72
Restoration, 68-70
Retro Scooters, 76-79
Rockers, 72, 81, 83, 112
Rockola, 16
Roman Holiday, 28-29
Ross, Dave, 81
Salsbury Motor Works, 14-20, 29-30, 54
Sarcastic Bastards, 104
Scarlett Fever, 107
Scoot Moab, 124-125
Scoot! Quarterly, 91
Scoot.net, 101
Scootabout, 17
Scootamoto, 13
Scooter Bible, The, 9
Scooter Boys, 87-88
Scooter Dolls, 55
Scooter Girls, 86-87
"Scooter Ice Age", 40, 44, 54
Scooter Racing Nationals, 86
Scooter Rage, 95, 115, 118-119
Scooter Shops, 33, 39, 40-46, 48, 77, 81, 91, 107, 138-139
Scooters Originali, 41
Scooters!, 9
Scooterworks, 40-42, 47, 64
Sears, Roebuck and Company, 30-35, 37, 38
Seattle Scooter Insanity, 115, 122
Secret Servix, 94
Secret Society, 95
Shriners, 29, 65
Sidecars, 17, 38, 74, 75
Slaughterhouse, 130-131
Sport Scooters, 57-58
Sputnik, 103
Sqream, 104
Stafford, Tim, 90
Teddy Boys, 111-112
Ten Year Lates, 102
Torre, Pierluigi, 23
Traditional Scooters, 47-48, 53, 61
Triumph, 27, 83
Tucson/Nogales Fall Classic, 91, 127
Twist & Play, 97
Twist-N-Go, 48, 50, 53
Unrestored Scooters, 66-67
Upstart, 98
Velocifero, 42-43, 46, 56
Vespa, is everywhere
Vespa Club of America, 46, 60, 65, 88, 89, 92, 98, 117, 136
Vespa Supershop, 41, 85
Vintage Scooters, 62-65
Volugrafo, 17
Websites, 138
Weddings, 108
Welbike, 17
WKRP in Cincinnati, 130
World War II, 16, 19, 20, 22, 26, 53
Wussys, 96
Yamaha, 28, 39, 40, 42, 46, 48, 50, 53, 54, 55, 56, 67
Zip Scoot, 16
Zundapp, 27

Bibliography

Brown, Gareth. *Scooter Boys*, 4th ed. Jobz for the Boyz, Ltd.: Essex, UK, 1996.

Calabrese, Omar, editor. *The Cult of Vespa*. Piaggio: Pontedera, Italy, 2001.

Dregni, Michael and Eric Dregni. *Scooters*. Motorbooks International: Osceola, WI, 1995.

Dregni, Michael and Eric Dregni. *Illustrated Motorscooter Buyer's Guide*. Motorbooks International: Osceola, WI, 1993.

Dregni, Eric. *Scooter Mania!* Motorbooks International: Osceola, WI, 1999.

Walker, Alastair. *Scooterama: Café Chic and Urban Cool*. Motorbooks International: Osceola, WI, 1999.

Photo and Illustration Credits

The publisher would like to thank the many individuals, associations, and scooter manufacturers who helped us gather images to make this a phenomenal book. Thanks!

©Dave McGrath, www.daveshoots.com, p: 1, 2, 54, 59, 71, 87, 121, 125, 127, 139
Robert Lawrence, p: 4
Dana Wilson, p: 6, 11, 79, 80, 81, 126,
Eric and Michael Dregni, p: 9, 16, 19, 20, 23, 31
Author's Collection, p: 10, 17, 24, 32, 34, 36 (Cushman), 41, 45, 51, 56, 64, 66, 82, 85, 91, 93, 94, 98, 101, 106, 107, 110, 112, 114, 115, 118, 120, 121, 124, 127, 130, 137, 139
©Genevieve Naylor / Corbis, p: 14
©Bettmann / Corbis, p: 18, 25, 26, 36 (couple)
©Negri, Brescia / Corbis, p: 22
Derek Lawrence / Jim Dillards' scooters, p: 28
Alex Zangeneh Azam, p: 29, 96, 97, 104, 122
©Hulton-Deutsch Collection / Corbis, p: 30, 84
©Glenn Reid Studio, www.reidstudio.com, p: 35, 37, 138
Robin & Chris Hedlund, p: 38, 116, 117, 128, 129
©Genuine Motor Company, p: 39, 49, 63
P. J. Chmiel / Scooterworks, p: 42, 43
©Italjet USA, p: 46
©Vespa USA, p: 47, 60, 63
©TN'G Scooters, p: 52
Craig Vetter, p: 53
The Scooter Dolls, p: 57
©Malguti USA, p: 58
©Aprilia USA, p: 59, 61
Minnesota Maxiscooter Club, p: 61, 106
©Piaggio USA, p: 62
©Bajaj USA, p: 63
Paul Taylor, p: 64
Ryan Basile, p: 65, 90, 91
Mike McWilliams, p: 67
Paul Kavinaugh, p: 68
Natalie Lachance, p: 69
Dean Wright, p: 70
©Gary Isaacs, p: 71, 108, 143
Damir Gusic, p: 72
Aidan Lawrence, p: 73, 94
Adam Baker, p: 73, 76
Michael Anhalt, p: 74
Nyle Schafhauser, p: 75
Josh Snow, p: 76
Richard Guile, p: 77
Dennis Memmott, p: 77
Dave McCabe, p: 78
***Scoot!* / Dave Ross** (Scooterist Trading Cards), p: 84, 86, 88, 89, 90, 91
Mykal Powell, p: 86
©2003 Peter Smakula, p: 88
Missi Kroge, p: 88
Chelsea Lahmers, p: 89
©Richard Black, p: 95, 102
Eric de Jong, p: 96
Oregon SC, p: 99
Paul Felix, p: 100
©Don Kilpatrick, www.donkilpatrick.com, p: 103
Sqream SC, p: 106
Hill City United SC, p: 106
Sarcastic Bastards SC, p: 106
Checkered Demons SC, p: 106
©Ivy Brooks, p: 109, 133
Birmingham Scooter Syndicate, p: 111
©John Stafford, p: 116, 130, 131, 135, 136, 139
©2004 Damon Landry / L Squared Studio, p: 119, 132, 134
Kieran Walsh, p: 136

DMS